MIDLAND TIMES

CONTENTS

Introduction	3
LMS Accidents and Breakdown Cranes	4-9
Express freight on the former GSWR route	10-17
My Trainspotting Odyssey (part 2)	18-23
The carriage of goods by rail	24-27
Midland Railway 'Flatiron'	28-29
The Beyer-Ljungstrom Turbine Locomotive	30-33
LMS motive power depot codes after 1935	34-41
The Midland Railway's Swansong	42-54
Collision at Chapel-en-le-Frith	55-57
Blackpool's Railways	58-69
49509 at Huddersfield	70
Far reaches of the LMS	71-78
46246 at Cheddington	79
The Platform End	80

Caledonian Railway McIntosh Class 711 0-6-0 'Standard Goods' No. 57473 at Perth MPD on 26th June 1956. Built at St. Rollox Works in February 1897 and entering service as CR No. 587, becoming No. 17473 a year after the 1923 Grouping. The loco had a long working life of 64 years and four months before withdrawal from service on 27th June 1961. PHOTO: DAVID ANDERSON © TRANSPORT TREASURY

© Images and design: The Transport Treasury 2024. Design and Text: Peter Sikes
ISBN: 978-1-913251-81-9
First published in 2024 by Transport Treasury Publishing Ltd., 16 Highworth Close, High Wycombe HP13 7PJ

The copyright holders hereby give notice that all rights to this work are reserved.
Aside from brief passages for the purpose of review, no part of this work may be reproduced, copied by electronic or other means, or otherwise stored in any information storage and retrieval system without written permission from the Publisher.
This includes the illustrations herein which shall remain the copyright of the copyright holder.

Copies of many of the images in MIDLAND TIMES are available for purchase/download.
In addition, the Transport Treasury Archive contains tens of thousands of other UK, Irish and some European railway photographs.

www.ttpublishing.co.uk or for editorial issues and contributions email MidlandTimes1884@gmail.com

Printed in England by Short Run Press Limited, Exeter.

INTRODUCTION

We hope you enjoy the fifth edition of Midland Times, with more articles and photographs from the LMS/BR(M) region.

Our first article *LMS Accidents and Breakdown Cranes* by Peter Tatlow discusses the poor safety record of the LMS), highlighting numerous accidents and the company's reluctance to adopt recommended safety technologies. Peter reflects on his personal interest in railways, influenced by the Harrow disaster, and his later career with British Railways.

Next, *Express freight on the former GSWR route* by Ian Lamb highlights the significance of freight operations in the early railway era, particularly on the Glasgow & South Western Railway (GSWR) route from Glasgow to Carlisle via Kilmarnock. It recounts a memorable freight journey in the 1950s, emphasising the challenges faced by train crews navigating harsh weather conditions and steep gradients.

Part Two of *My Trainspotting Odyssey* by Geoff Courtney, is a nostalgic recount of the author's trainspotting experiences on the London Midland Region in the late 1950s and early 1960s. It details visits to various stations as the author observes the transition from steam to diesel traction during that period.

The carriage of goods by rail, details the impact of the Rail and Canal Traffic Act of 1854 and subsequent legislation on British railways, especially regarding freight transport. The author explains the inefficiencies and financial difficulties caused, exacerbated by further legislation and the rise of road haulage in the 1920s.

In *The Beyer-Ljungstrom Turbine Locomotive*, David Cullen explores the development and challenges of steam locomotive technology in Britain during the 1920s. Specifically, it details the Beyer-Ljungstrom Turbine Locomotive, an experimental design using steam turbines, commissioned in 1926 by the LMS. The article contrasts the Beyer-Ljungstrom's limited success with its pioneering influence in the later adoption of turbine technology elsewhere.

LMS motive power depot codes after 1935 by David Young explores the evolution of engine shed code numbers used by the LMS and later by the London Midland Region of BR after nationalisation in 1948. Examples of short-lived codes and their subsequent modifications are also discussed, illustrating the complexities of railway management and organisation.

The Midland Railway's Swansong by Philip Hellawell provides a comprehensive history of the Midland Railway, tracing its evolution from a local operation to a major national railway company. It explores key milestones such as strategic mergers which propelled the MR's growth, led by figures like William Stenson and George Hudson.

Collision at Chapel-en-le-Frith recounts the tragic incident involving John Axon, a dedicated engine driver from Stockport, who lost his life while attempting to halt a runaway freight train on a steep gradient near Chapel-en-le-Frith, Derbyshire, on 9th February, 1957.

Blackpool's Railways looks at the historical development of railways connecting Preston to the Fylde Peninsula, focusing on routes to Fleetwood and Blackpool established in the mid-19th century. It describes how these railways fostered tourism and economic growth but later faced decline with the advent of cars and buses, and shows the efforts made by local railway enthusiasts to preserve parts of these lines as heritage railways.

Our closing article, *Far reaches of the LMS*, by Alan Postlethwaite presents a series of photos capturing BR(M) operations from 1959 to 1967, taken during various trips. Highlights include photographs from the Highlands, taken during the author's honeymoon in Inverness, featuring trips to Aviemore and Kyle of Lochalsh.

To make sure you never miss an issue of **Midland Times** why not sign up to our subscription service? For details visit **www.ttpublishing.co.uk**, email **admin@ttpublishing.co.uk** or call us on **01494 708939**.

PETER SIKES, EDITOR, MIDLAND TIMES
email: midlandtimes1884@gmail.com

FRONT COVER (AND ABOVE): Fowler Class 3F 0-6-0T No. 47658 pictured at Crewe Works on 13th October 1963. One of 90 locos of the class that was built by William Beardmore and Company, Dalmuir, Glasgow. The company was better known as a shipbuilder that diversified into the manufacture of steam locomotives. 47658 was built in 1929, lasting in service until October 1966 and scrapped by Cohens of Kettering during February 1967. PHOTO: BALHAM COLLECTION © TRANSPORT TREASURY

LMS ACCIDENTS AND BREAKDOWN CRANES

by Peter Tatlow, President of the LMS Society

Accidents on the LMS

The LMS did not have a particularly good record with respect to accidents; inheriting the dubious honour of including the site of the worst railway accident ever in the United Kingdom, at Quintinshill on the erstwhile Caledonian Railway, when 230 people lost their lives. Since the beginning of 1923 up to 1942, the LMS was responsible for 192 (45%) of those killed and 996 (43%) of the injured among the four grouping companies. Its nearest rival, the LNER, accounted for 35/36%, Southern 8/14%, whilst the GW was down at 9/6%.[1] Along with two of the other companies, the LMS had been hesitant to install Automatic Train Control/Automatic Warning System, for so long persuasively advocated by the Railway Inspectorate.

On 3rd November 1924 at Lytham the derailment at speed of an express businessmen's train from Manchester and Liverpool heading for Blackpool had resulted in fourteen fatalities and ten seriously injured. Four years later the overnight mail train from Derby to Bristol overran a signal at Charfield, colliding with the engine of a freight train setting back into a siding, and was deflected into the path of a further freight train passing on the adjacent track. This resulted in debris piling up against an overbridge, only for fire fuelled by gas tanks to break out. Fifteen persons lost their lives and 24 were injured.

1929 was barely a week old, and a not dissimilar accident occurred at Ashchurch, when a Compound with a fourteen-coach express train, having overrun several signals at danger, ploughed into a freight train setting back over a trailing crossover, resulting in three deaths and 24 injured. Year upon year, further serious accidents took place at Leighton Buzzard (3 killed/15 injured), Great Bridgeford (4/51), Little Salkeld (nil/27) and Port Eglington, Glasgow (6/55). Eleven passengers lost their lives at Winwick Junction, Warrington on 28th September 1934 and 72 were injured when two passenger trains collided leading to telescoping of some coaches. Grave incidents continued to happen during the war, such as Oakley Jct. (2/41), Rutherglen (2/15), Elderslie (1/108), Bletchley (4/41), Winwick Junction again (1/nil), Holme Chapel (9/10) and Eccles (23/56).[2]

Despite the cessation of hostilities in 1945, the toll continued to mount with serious accidents at Bourne End (43 killed/124 injured), Lichfield (20/20) and Polesworth (5/19). Even following nationalisation from 1948 accidents occurred on ex-LMS territory at Winsford (5/19), Weedon (15/36), Blea Moor (nil/34) and culminating in the disaster at Harrow & Wealdstone in the autumn of 1952, resulting in the deaths of 112 people, the second most serious accident in the country.

The country was shocked by the disastrous double collision which occurred during the morning rush hour at Harrow & Wealdstone station on 8th October 1952. The two trains concerned in the initial rear end collision were the 7.31am Up local passenger train from Tring to Euston, comprising nine non-corridor bogie coaches hauled by a Fowler 2-6-4 tank engine, No. 42389, and again the 8.15pm Up overnight express passenger train from Perth to Euston, made up of eleven bogie vehicles hauled by Stanier Pacific No. 46242 *City of Glasgow*. By great misfortune, within seconds, a third train ran into the wreckage of the first collision. This was the 8.00am Down express from Euston with portions for Liverpool and Manchester, consisting in total of fifteen bogie vehicles double-headed, by 4-6-0 Jubilee class No. 45637 *Windward Islands* piloting the recently rebuilt former 4-6-2 Turbo-locomotive No 46202 *Princess Anne*.[3]

By 1952, I was travelling daily to school by Southern electric train and, on seeing the headlines on the news-stand of the railway disaster at Harrow & Wealdstone station the previous day, purchased my own copy of the Daily Telegraph. A growing interest in railways as a whole, Hornby-Dublo model trains and the writings of the Reverend Edward Beal in particular, had already made me aware of the work of steam breakdown cranes. Here in the paper were vivid pictures of these beasts, fulfilling their calling, by removing wreckage to enable the injured to be released, the bodies of the dead recovered and in due course helping to clear and restore the line to working order.

Later, following three years military service in the Royal Engineers, I joined the Chief Civil Engineer's Department of the Southern Region of British Railways. Yet despite the collapse of the flyover at St John's, Lewisham within two months, I was fortunately never personally involved in an event as traumatic as Harrow & Wealdstone, but did become familiar with the use of breakdown cranes for bridge works and re-railing errant vehicles.[4]

References
1. *MOT Accident Returns (PRO, ref RAIL1053/109-114).*
2. *Earnshaw A, Trains in Trouble, Atlantic, 1996.*
3. *Tatlow P, Harrow & Wealdstone 50 years on, Clearing up the aftermath, Oakwood Press, 2002 & 2008.*
4. *Tatlow P, St. John's, Lewisham, 50 years on, Restoring traffic, Oakwood Press, 2007.*

3RD NOVEMBER 1931 • Due possibly to appalling weather conditions, a freight train ran into a train of empty coaching stock just north of Tring station. LMS 2-6-0 Crab No. 13153, hauling a freight train, had been built the previous year. The also then nearly new Craven Bros. 36-ton steam breakdown crane, probably from Rugby, is seen here during the clearing up operation. PHOTO: HERBERT MORRIS, AUTHOR'S COLLECTION

21ST SEPTEMBER 1951 • With the continuing effects of delayed maintenance following the war, when the expedient of swapping round the wheel and axle of the bogie of a locomotive met a length of track overdue for renewal, disaster ensued at Weedon. No. 6207 PRINCESS ARTHUR OF CONNAUGHT with the 8.20am 15-coach train from Liverpool to Euston, having passed through Stowe Tunnel south of Weedon, derailed. The rapid deceleration of the locomotive, as it fell on one side, caused seven of the following coaches to concertina into a zigzag pattern. The driver and 14 passengers died in the ensuing mêlée. Here several ambulances and fire engines, together with Police and other cars, are in attendance with their crews attempting to rescue the injured and extract the dead. PHOTO: UNATTRIBUTED, AUTHOR'S COLLECTION

13TH MARCH 1935 • The 4.55pm express meat train on its way from Alexandra Dock (Liverpool) to Broad Street (London) stopped in section on the busy West Coast Main Line at King's Langley to attend to a defective brake pipe. Due to the signalman's inattention, this quickly led to a rear end collision by the following 5.50pm Stafford to Euston milk train. The resulting obstruction of the adjacent track by the debris was promptly run into by Patriot Class No. 5511 with the 10.30pm Camden to Holyhead partially fitted freight train, only to be followed by a 70 wagon coal train. No. 5511 is seen here with wagons piled up beside and beyond. The tops of two crane jibs are apparent on the right-hand side starting to tackle the task of clearing away the debris while officials and others stand by watching. The style of their headwear was an indication of the status of the wearer. PHOTO: TOPICAL PRESS, AUTHOR'S COLLECTION

8TH OCTOBER 1952 • By early afternoon at Harrow & Wealdstone the 30-ton Cowans Sheldon crane from Rugby reached over the mountain of debris as rescuers continue to search for bodies and hopefully a last remaining survivor. Locomotives Nos. 45637 and 46202 of the Liverpool train lie on their sides in the foreground on the Up Electric line, having been deflected across platforms 2 and 3 by the obstruction caused by the engine of the overnight sleeper train from Perth.
PHOTO:
METROPOLITAN POLICE MUSEUM, AUTHOR'S COLLECTION

LMS Breakdown Cranes

With the grouping in 1923, the LMS inherited from its constituent companies twenty-five steam breakdown cranes of varying size, allocated to the Motive Power Department. Many were of low-capacity dating back to the last decade of the previous century. Nonetheless, it did include five relatively modern 35/36-ton cranes, including one with relieving bogies, together with two 25-ton and three 20-ton not quite so up-to-date cranes.[5]

As the company set about building increasing numbers of larger and heavier engines for use on the system as a whole, the smaller cranes were soon out-classed, resulting in the need to bring in a bigger crane from further afield. To meet this challenge, in 1930 the LMS ordered six 36-ton cranes from three different builders. The three from Cowans and two from Craven and a single crane from Ransomes & Rapier, were all equipped with relieving bogies to spread the load while in train formation.

The arrival of William Stanier led to the introduction in 1933 of even bigger engines. A further 520 engines of over 72 tons entered service between the beginning of 1931 and the end of 1938, the Princess Coronation Pacifics weighing in at 108 tons without tender.

As with so much of the LMS during this period, it seems that a review of the Motive Power Department's breakdown arrangements took place around 1937/8. The Cowans and Craven 36-ton cranes had been designed to handle a maximum load of 36 tons from the minimum radius of 18 feet out to a maximum of 25 or 24 feet. The possibility that greater loads at reduced radii would be lifted under emergency conditions was acknowledged by the LMS in their strengthening of this type to 50 tons at minimum radius.

Before the last war, out of the LMS's stock of now 31 steam breakdown cranes, just less than half were of 15-ton capacity dating from the end of the 19th century, but nonetheless handy machines for the more mundane incident of a derailed wagon or two in a yard. The Company therefore set about developing a more modern counterpart of 30-ton capacity, with a long jib and yet uninhibited by limitations on route availability and without the encumbrance of relieving bogies or removable counterweights, etc. Following delivery of a prototype in 1941, ten more were constructed at the height of the war, the order being split equally between Cowans and Rapier. As well as the new cranes and match wagons, the provision of additional staff riding and tool vans with associated equipment was put in hand. Once delivered,

BELOW: Cowans Sheldon 15-ton Midland Railway No. 26 from Saltley and the 36-ton Ransome & Rapier steam crane from Derby in the process of recovering Midland 'Flatiron' 0-6-4T No. 2015, in the vicinity of Halesowen Junction in 1921. The Midland Railway had since 1915 maintained this high-capacity crane (36T) at the centre of its system, Derby, to act in support of a number of lesser capacity cranes (15T) stationed strategically around the rest of the network. PHOTO: W. L. GOOD, AUTHOR'S COLLECTION

the old cranes they were intended to replace were nonetheless retained in order to strengthen the breakdown resources and deal with possible emergencies arising out of enemy action.

Arising from the Harrow & Wealdstone accident, British Railways in due course acquired a dozen 75-ton and ten 30-ton breakdown cranes from Cowans Sheldon in 1959-1964, two of each were diesel powered for the Southern Region. From 1977 the ten steam 75-ton cranes were converted to diesel propulsion.

References
5. *Tatlow P, Railway breakdown cranes, Vol. 1 & 2, Noodle Books, 2012 & 2013.*

This photograph was taken at Newton Heath, probably on the occasion of the delivery of a Craven Bros. crane circa 1931. On the left is a Cowans Sheldon 20-ton crane supplied to the L&YR in 1902, which is about to be replaced by one of a pair of brand new 36-ton cranes for the LMS from the Craven Bros. stable. Bearing in mind that the latter is positioned further from the camera, one can gain some idea of their relative size. Whilst the carriages of both are mounted on four axles, in addition the Craven crane has a detachable 4-wheel relieving bogie at each end to distribute the greater weight while in train formation.
PHOTO: CRAVEN BROS., AUTHOR'S COLLECTION

During World War 2, the LMS received a total of eleven modern 30-ton cranes to replace outdated cranes on their system. The prototype from Cowans Sheldon, showing no signs yet of having been numbered RS1066/30, is still paired with its first match wagon, ex-MR 30-ton bogie rail wagon No. 117148. Some remnants of lining are visible on the crane, but excluding the match wagon.
PHOTO: LMS, AUTHOR'S COLLECTION

Tool vans were created by plating over the sides of redundant bogie coaches, providing two sets of wide sliding doors through which to pass items of heavy equipment stored inside. LMS No. 284651 was an ex-Caledonian 57-foot coach labelled for Inverness, although it is thought to have ended up elsewhere. This photograph displays a short length of rail hung on hooks below the solebar and the steps included to afford access from the lineside. A second vehicle would be provided as a riding van for the breakdown gang.
PHOTO: LMS, AUTHOR'S COLLECTION

LMS 30-ton Ransomes & Rapier crane No. RS 1067/30, the first of the batch of five from Rapier, soon after delivery in 1942 when based at Wellingborough. The carriage framework and elements of the superstructure, but not the jib, are lined out. This was omitted before the delivery of other cranes of the order was complete.
PHOTO: LMS, AUTHOR'S COLLECTION

The driver looks up as his Ransomes & Rapier 30-ton crane No. RS1071/30 based at Polmadie, Glasgow prepares to assist ex-LNER 45-ton No. RS1058/45 (out of view) in the dismantling of the redundant flyover at Stevenson in 1949.
PHOTO: J. TEMPLETON, AUTHOR'S COLLECTION

EXPRESS FREIGHT ON THE FORMER GSWR ROUTE
by Ian Lamb

An express freight crosses Ballochmyle Viaduct on 15th September 1962. PHOTO: W. A. C. SMITH © TRANSPORT TREASURY

The glamour of passenger train operation is all very well, but freight – certainly in the early years of the railway – was the lifeblood for the existence of many lines. Such services were provided regularly and reliably over a ribbon of steel stretching through town and country. In relatively hilly areas like that of the former Glasgow & South Western Railway's route from Glasgow to Carlisle via Kilmarnock, the route could be particularly prone to the natural elements, even more so in winter conditions.

I was still very much in 'short trousers' when I came across a reference to the following remarkable journey and stuck it in my railway scrap book, and I am now willing to share the experience with you.

This particular early Spring day in the 1950s may well have been devoid of snow, but when a crosswind rises, for all the heat from the engine's fire, it can be like an ice-box on the footplate, especially if you are shovelling coal. Prior to allocation of a journey with a freight train from Kilmarnock to Carlisle, the locomotive and crew commenced their duty with a passenger train from Glasgow St. Enoch station.

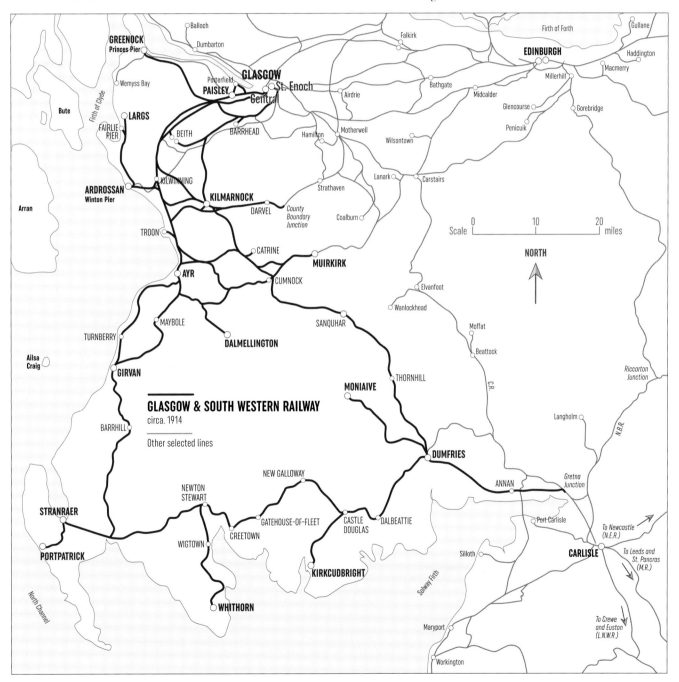

19TH DECEMBER 1953 • Not long out of St. Enoch is the closed Gorbals station, seen here as the 10.40am train from Kilmarnock rushes through the winter gloom. PHOTO: W. A. C. SMITH © TRANSPORT TREASURY

Proceeding from 67A Corkerhill – the GSWR's principal shed – the Stanier 4-6-0 engine and men connected with the moderately loaded 5.3pm passenger train from St. Enoch to Kilmarnock by way of Barrhead over the severely graded Shilford Bank, dreaded by drivers of the early locomotives and still treated with respect by their successors. However, this 'Black 5' experiences no difficulty, and scheduled time was easily maintained.

At the time of the grouping the Glasgow & South Western Railway was the fourth largest railway company in Scotland, and the tracks connecting Kilmarnock were busy with both passengers and freight. The many coal pits of the surrounding area had their output moved to the coast for shipment and to various inland destinations as the Machine Age developed. In its own right it was the manufacturing base for the GSWR industry, but also blossomed in general as it became possible to reach more and more customers.

On arrival at Kilmarnock the passengers alight, sundry parcels and packages are unloaded, and the train is deposited in a siding. The engine is run alongside a water column and when the water tank has been refilled the 'Black 5' proceeds to the goods yard.

TOP LEFT: Stanier Class 5MT 4-6-0 No. 44672 departs Glasgow St. Enoch in this undated image, while Standard 4MT 2-6-4T No. 80047 waits for its departure time.
PHOTO: NEVILLE STEAD COLLECTION © TRANSPORT TREASURY

BOTTOM LEFT: 22ND OCTOBER 1955 • Fowler Class 2P 4-4-0 No. 40618 is the pilot engine for Fairburn Class 4MT 2-6-4T No. 42191 as they prepare to double-head the 10.40a.m. 'Empress Voyager' Liner Special to Greenock Princes Pier from Glasgow St. Enoch.
PHOTO: W. A. C. SMITH © TRANSPORT TREASURY

Here marshalling of the train for the south is being completed with traffic brought in by the last of the 'feeder' trains, timed to arrive a few minutes before scheduled departure. This arrangement gives traders the maximum time to prepare their goods and ensures that they are transported to distant destinations without delay.

The scene at the marshalling yard is one of bustling activity with shunters coupling or uncoupling wagons and the shunting engine hurrying to and fro as the driver responds to the signals. Supervising the scene is the yard foreman, glancing occasionally at his watch as the appointed time draws near. Under the new British Railways plan of the period, all freight trains were to be fitted with continuous braked vehicles. In this instance it was a case of a 'mixed bag'.

"That's 38 on, 20 fitters" says the guard to the driver. "Just draw away quietly till I get on my van." The '20 fitters' are vehicles fitted with the continuous brake equipment and coupled to the engine to give the driver control of adequate brake power for this express goods train. Turning to his firemen the driver says, "You're going to be busy, son, we've got a full load on." The fireman cheerfully replies "Carry on Mac, I'm ready."

The upper quadrant 'home' signal is cleared to allow the train on to the main line, then the regulator is opened slightly to start the wheels turning, and in all probability they will continue to turn without stopping throughout the 90 miles to Carlisle. Looking back from the driver's position, the train seems to disappear as the engine turns into a right hand circumference, then one by one the wagons and vans reappear as they come out of the curve on to a straight stretch of line.

26th December 1958 • Black 5 4-6-0 No. 44668 with the 1.53pm Glasgow St. Enoch to Carlisle train passes the junction with the closed Barrhead Central branch.
Photo: Neville Stead Collection © Transport Treasury

21st August 1963 • Black Five 4-6-0 No. 44785 double heads with Peppercorn Class A1 4-6-2 No. 60131 Osprey through Kilmarnock station on the 6.38pm car sleeper service (1M90) from Glasgow St. Enoch to London Marylebone.
Photo: W. A. C. Smith © Transport Treasury

This journey begins on a series of slight falling gradients so this freight train gets away to a good start with the driver forcing the pace before reaching the heavy climb which begins less than two miles away. On reaching the 1 in 100 gradient, speed falls steadily until the engine is slogging away at a steady pace. At short intervals the fireman puts a few shovelfuls of coal on the fire to maintain steam at full pressure.

To prevent a further fall in speed as the climb continues, the driver occasionally makes careful adjustments of the controls. After some fifteen minutes of this procedure Mossgiel Tunnel shows up ahead. Steadily the entrance portal looms, growing larger until this freight train passes through it into the darkness. A good head of steam plus smoke from the engine's chimney strikes first upwards to the roof of the tunnel, then spreads swiftly down its walls.

Now in the heart of Burns Country, for the land above this tunnel was once farmed by the poet and his brother, and here he wrote some of his best work. A mile beyond the tunnel the train passes through Mauchline (again richly associated with Robert Burns). Further on still is the magnificent Ballochmyle Viaduct which crosses the River Ayr. Only a fleeting glance can be had of this strikingly beautiful scene with tree lined banks falling steeply to the winding river 169 feet below.

Still climbing upwards, this freight train soon emerges into the extensive coal bearing area centred around Cumnock. The scenery changes to bare moor and hill, a far cry from the same countryside many millions of years ago, when thick forests, which became the coal measures of the 1960s, grew in humid tropical valleys. A rather difficult exercise this in present-day conditions, but particularly so when the country is in the grip of winter and snow lies everywhere!

Beyond New Cumnock the water trough built between the rails is reached and the fireman crosses the footplate to dip the water pick-up attached to the underside of the tender. Some 1,400 gallons of water used since the tank was filled at Kilmarnock are quickly replaced, and with a full tank once more there is enough water to take the train to Carlisle 70 miles away. To be certain, the water pick-up will probably be lowered again when reaching Floriston trough 61 miles distant.

Two miles after taking water the beginning of a series of falling gradients is reached, which, with the exception of short stretches of level and the inclines to Ardoch and Drumlanrig tunnel, form the route to Dumfries nearly 32 miles ahead. For some twelve of these miles the line runs through the Nith Valley, presenting some wonderful views of the river splashing angrily over its boulder strewn bed, and sometimes flowing smoothly in

Ballochmyle Viaduct was one of the greatest engineering achievements of its time – the world's largest masonry span arch. Work started in March 1846 and was completed two years later before the railway itself became operable in 1850. It still claims to be the UK's highest railway viaduct. The bridge became a star in the film 'Mission Impossible' when the famous scene on the top of the train was filmed while the train crossed over the viaduct. PHOTO: W. A. C. SMITH © TRANSPORT TREASURY

deep reaches between tree-lined banks. A very pleasing strip of country and well worth visiting.

On this long downhill stretch, the controls are adjusted to admit just a whiff of steam to the cylinders to balance the swiftly moving pistons. The engine runs easily, and the 38 vehicles behind sway slightly from side to side, their separate movements making the train appear to wriggle like a giant caterpillar.

After Dumfries, the only obstacle of note is the 1 in 150 Ruthwell bank, but this 'Black 5' takes the grade without difficulty and speeds on towards Annan and Gretna Junction; passing the little village of Gretna Green, historic scene of many runaway marriages. Up to this point the run has been entirely free from signal checks and now looking ahead to where the junction signals will appear wondering what message they will convey. Again they are at 'clear' and the train swings round the bend to race on through Floriston and Rockcliffe. Then the regulator is closed and the train coasts at gradually decreasing speed until it comes quietly to a stop in Kingmoor Yard.

The 'Black 5' class were not only the general workhorses of the LMS and their constituencies, but the company also had a 'common user' policy so it was not unusual to 'spot' a locomotive far from its home shed. Consequently, as far as this Glasgow St. Enoch–Kilmarnock–Carlisle journey was concerned, No. 45373, allocated to Crewe (5A) on Dumfries MPD (68B), was no exception.

27TH JULY 1963 • Carstairs (64D) 'Black 5' 4-6-0 No. 45010 with an 'Up' freight near Kirkconnel riding high above the main road and the river in the Nith Valley. PHOTO: W. A. C. SMITH © TRANSPORT TREASURY

15TH APRIL 1963 • Jubilee' 4-6-0 No. 45659 DRAKE leaves Dumfries with the 8.30am relief train from Glasgow St. Enoch to London St. Pancras, passing 'Black 5' No. 45373 which is in between duties and resting on shed. PHOTO: W. A. C. SMITH © TRANSPORT TREASURY

AUGUST 1964 • A Stanier Jubilee accelerates northwards out of Dumfries station with a St. Enoch bound express. The track to the left heads cross-country to Lockerbie, whilst the rails to the right are the main route to Stranraer. PHOTO: W. A. C. SMITH © TRANSPORT TREASURY

MY TRAINSPOTTING ODYSSEY (PART 2)
by Geoff Courtney

Having launched my late-1950s/early 1960s London Midland Region trainspotting odyssey in the previous issue of Midland Times with logs at Oxenholme and Rhyl, I complete my two-part series with my notes from four other stations, starting with Bedford Midland Road on Monday 7th March 1960.

I was there for just over an hour, a short stay which gave me no time to visit the locomotive depot there (14E). It did, however, give me time to log the Midland Pullman on an Up trial run, four months before the service was launched between St. Pancras and Manchester Central. These 90mph six-car diesel-electric units were built by Metropolitan-Cammell, and cut the journey time between the two cities by 35 minutes.

Steam, though, was the predominant motive power that morning, and included Patriot No. 45536 *Private W. Wood, V.C.* on the Down 'Thames-Clyde Express,' Jubilee No. 45589 *Gwalior* on a Down Manchester working, and classmate No. 45636 *Uganda* on a Down semi-fast to Leicester.

An Up double-header on what I logged as a semi-fast comprised Black Five No. 44817 and Standard Class 5MT No. 73135, and freight haulage included Class 9F Nos. 92078 and 92116, each on Up trains, while two Down goods trains were hauled by Class 8F Nos. 48619 and 48734.

Five months later, on Saturday 6th August, it was back to the Midland Region, for a seven-hour visit to the WCML station of Lichfield Trent Valley, which brought home to me the mixture of steam and diesel traction which was so much part of the railway scene at that time.

To illustrate this, I will begin by listing chronologically the named trains I noted over those seven hours and the locomotives that pulled them. In the first hour I logged the Down 'Lancastrian' headed by Britannia No. 70043 *Lord Kitchener*, Class 40 D217 on the Up 'Emerald Isle Express,' Princess Coronation No. 46247 *City of Liverpool* on the Down 'Royal Scot,' and three more Class 40 diesels, D226 in charge of the Up 'Mancunian,' D230 on the Up 'Merseyside Express,' and D220 on the Down 'Comet.'

The second hour started with a return to steam, Royal Scot No. 46124 *London Scottish* on the Down 'Empress Voyager' boat train that connected with Canadian Pacific sailings at Liverpool on certain days of the week, and continued with D222 on the

Black Five No. 45187, which was logged double-heading a Down express with classmate No. 45230 at Lichfield Trent Valley on 6th August 1960, picks up a number of passengers, possibly on a rail tour, at an unidentified station, believed to be between Bletchley and Bedford on the Oxford-Cambridge 'Varsity Line.' PHOTO: STEPHEN SUMMERSON © THE TRANSPORT TREASURY

The fireman of rebuilt Patriot No. 45540 SIR ROBERT TURNBULL watches two men behind the 4-6-0, one of whom appears to be wearing a suit and trilby and carrying a flag, outside Camden shed (1B), two miles from Euston, on 21st September 1958. Geoff Courtney found this north-west London depot difficult to 'bunk' and so frequently resorted to travelling past by train to note the locomotives on shed.
PHOTO: R. C. RILEY © THE TRANSPORT TREASURY

Down 'Manxman,' after which the afternoon progressed to Royal Scot No. 46170 *British Legion* heading the Down 'Welshman,' Princess Coronation No. 46252 *City of Leicester* hauling the Down 'Lakes Express,' and D229 on the Down 'Red Rose.'

As the afternoon edged towards early evening, steam nicked it 4-2, thanks to Princess Coronation No. 46238 *City of Carlisle* on the Down 'Mid-Day Scot,' Royal Scot No. 46114 *Coldstream Guardsman* in charge of the Up 'Lakes Express,' classmate No. 46125 *3rd Carabinier* on the Up 'Welshman,' and another Princess Coronation, No. 46248 *City of Leeds*, on the Up 'Royal Scot.' Diesel's response consisted of Class 44 'Peak' D9 *Snowdon* in charge of the Up 'Manxman' and D268 with the Up 'Lancastrian.'

Before dwelling on other steam workings, I must record that Class 40 diesels were at the helm of 11 further expresses, a second Class 44, D7 *Ingleborough*, made an appearance on a Down Liverpool train, and beating the drum for slightly older diesel-electric traction was December 1950-built 1Co-Co1 No. 10201 on an Up Liverpool express.

So, what of steam on non-titled trains? Royal Scots and Jubilees shared 18 between them, five Patriots featured, and two Princess Royal Pacifics, one of which, No. 46204 *Princess Louise*, was (according to my notes) on nothing grander than a semi-fast to Crewe. The only Princess Coronation on a non-titled express was No. 46245 *City of London* on an Up working.

Black Fives inevitably also had a hand in proceedings, including Nos. 45187 and 45230 double-heading a Down relief express, while during the day I logged just two freight trains, worked by Class 4F 0-6-0 No. 44332 and Class 8F 2-8-0 No. 48152.

After seven hours it was time for me to leave, and I wasn't to record any further Midland Region logs until the following year of 1961, with short spells at two more stations on the WCML, South Hampstead on 26th April and Rugby Midland on 23rd May.

The former is a suburban station two miles out of Euston, and it was a journey I frequently undertook in order to log locomotives as we passed Camden shed (1B), which I recall was a difficult depot to 'bunk' even for us fearless trainspotters.

I was there for just 40 minutes, while I waited for the train to take me back to Euston, but that was long enough to note three titled expresses comprising the Up 'Ulster Express' headed by Class 40 D308, the Down 'Red Rose' with Princess Coronation No. 46240 *City of Coventry* in charge, and another Class 40, D303,

With steam to spare, Jubilee No. 45636 UGANDA is impatient to depart from St. Pancras with 'The Waverley' to Edinburgh on 23rd September 1959. This 4-6-0 was logged by Geoff at Bedford Midland Road on the same line heading a semi-fast to Leicester on 7th March the following year.
PHOTO: JAMES HARROLD © THE TRANSPORT TREASURY

pulling the Up 'Shamrock.' I also recorded another Princess Coronation, No. 46234 *Duchess of Abercorn*, on what I logged as an Up Wolverhampton semi-fast, and two diesel classes that entered the log on various workings were class 20 (D8008) and Class 24 (D5032, D5135, and D5141).

The main line into Marylebone passes over the WCML at South Hampstead station, and during my visit I noted, among other locomotives, Neasden (14D) allocated trio Class 4 No. 42082 and Standard classes 5MT No. 73010 and 4MT No. 76037 all running light, and Black Five No. 45217 on a Nottingham express.

A month later, on 23rd May, I was back on the WCML, this time at Rugby Midland, and during the early evening stay of two hours I was to be reminded, if a reminder was needed, that the heyday of steam on the Midland Region was coming to an end.

The visit started well enough, with Brit No. 70032 *Tennyson* on an Up express, but six minutes later the bubble burst when D330 came through on a similar working. The Class 40 scene was set, and I was to record no fewer than ten further members of the class, including D341 on the Down 'Caledonian,' D306 on the Down 'Shamrock,' and D328 on the Up 'Lancastrian.'

Princess Coronation No. 46240 *City of Coventry*, Royal Scot No. 46161 *King's Own*, and Jubilee No. 45634 *Trinidad*, did help to balance the books a tad on Up and Down expresses, aided and abetted by another Jubilee, No. 45737 *Atlas*, running light, and seven Black Fives on mainly semi-fast workings, but it was an inescapable realisation for even the most ardent steam enthusiast that a new era of motive power was not only dawning, but was in full flight.

With a career in journalism, which I am still fortunate to enjoy 63 years later, beckoning me to pastures new, my Midland Region trainspotting days were close to an end. Although my local line was the former GER route out of Liverpool Street, frequent visits to London and other parts of the country had resulted in me 'copping' the entire Princess Royal class, all but four of the Princess Coronations, all but five of the Royal Scots, and a majority of the Jubilees and Patriots.

And while Gresley's A4s were surely my favourite engines, I still reckon William Stanier's non-streamlined Princess Coronations were the most handsome of all the 'Big Four' express locomotives.

TOP LEFT: No. 46124 LONDON SCOTTISH creates an effective smokescreen at Rugby on 5th February 1953. Geoff noted this Royal Scot at Lichfield Trent Valley on 6th August 1960, heading the Down 'Empress Voyager' boat train, which connected with Canadian Pacific sailings at Liverpool. PHOTO: ERIC SAWFORD © THE TRANSPORT TREASURY

TOP RIGHT: No. 46238 CITY OF CARLISLE takes on water at Newbold Troughs north of Rugby with an Up express in April 1958. On 6th August 1960, the Pacific was noted by Geoff at Lichfield Trent Valley en route to Glasgow on one of the WCML's flagship expresses, 'The Mid-Day Scot.' PHOTO: ERIC SAWFORD © THE TRANSPORT TREASURY

MIDDLE: In an archetypal WCML scene, Stanier Pacific No. 46234 DUCHESS OF ABERCORN leaves the 355-yard Northchurch Tunnel at Berkhamsted, Hertfordshire, with the Up 'Caledonian' in September 1960. Geoff logged this locomotive at South Hampstead on 26th April the following year on a rather lower-profile train, an Up semi-fast. PHOTO: JIM FLINT AND JIM HARBART COLLECTION © THE TRANSPORT TREASURY

ABOVE: With a long rake of carriages behind the drawbar, Princess Coronation No. 46247 CITY OF LIVERPOOL heads through Rugby with the 'Royal Scot' on 29th May 1954. Geoff noted this Pacific at Lichfield Trent Valley on a Down working of this top-link express on 6th August 1960.
PHOTO: ERIC SAWFORD © THE TRANSPORT TREASURY

TOP: Princess Coronation No. 46251 CITY OF NOTTINGHAM was a member of one of the most glamorous and admired classes of steam locomotives within the BR stable, but its run-down and almost derelict surroundings here at Camden shed (1B) in July 1964 are the very antithesis of such a description. This itinerant Pacific moved home no fewer than 21 times in the 20 years between being delivered new to Crewe North shed in June 1944 and its withdrawal in September 1964, two months after this photograph was taken. Camden was one of Geoff Courtney's favourite London depots, despite his inability to 'bunk' the shed.
PHOTO: JIM FLINT AND JIM HARBART COLLECTION
© THE TRANSPORT TREASURY

CENTRE: Britannia No. 70032 TENNYSON rests between duties at its home WCML shed of Willesden (1A) in the summer of 1961. Geoff had logged the Standard Pacific earlier in the year, on 23rd May, heading a Euston-bound express at Rugby Midland.
PHOTO: DON MATTHEWS
© THE TRANSPORT TREASURY

LEFT: With its sights set firmly on its Manchester home of Longsight (9A), Britannia No. 70043 LORD KITCHENER passes through Carpenders Park, south of Watford, on 26th April 1958 with a Down express to the city. This 1953-built Pacific was noted by Geoff at Lichfield Trent Valley on 6th August 1960 also heading for Manchester, with the Down 'Lancastrian.'
PHOTO: AS D22-2
© THE TRANSPORT TREASURY

ABOVE: Class 40 D236 provides a powerful message that the steam era is coming to a close as it passes through Crewe with the Up 'Merseyside Express' on 14th April 1960, four months before Geoff Courtney logged classmate D230 on the same working at Lichfield Trent Valley on 6th August. PHOTO: BEN BROOKSBANK/CREATIVE COMMONS

BELOW: The Midland Pullman is the centre of attention as it passes the 1911-built Leicester North signal box on the Midland Main Line in 1960. This six-car diesel-electric train was noted at Bedford Midland Road on trial on 7th March that year. PHOTO: © THE TRANSPORT TREASURY

THE CARRIAGE OF GOODS BY RAIL

The Rail and Canal Traffic Act of 1854 was passed to maintain a degree of competitiveness between rail and canal systems. Under this Act the railways were obliged to carry any and all goods offered to them, other than where these might damage the rolling stock used or where they were physically too large for the bridges, tunnels and passing traffic on the line. Examples of the goods that could be refused would be materials such as tar coated stone chippings (tarmac) or lime, and a firm might prefer to use its own rolling stock to avoid contamination of its product by other cargoes, an example of this would be salt. Following this act and other subsequent legislation the term 'common carriers' came into use, referring to this obligation to carry goods offered no matter how small the consignment. The 1854 Act also introduced a government regulated national system of charges for freight moved by rail, based on the weight and value of goods carried.

These two factors, the obligation as common carriers and the Government regulated charges structure, were to become the two biggest obstacles to the railway's ability to compete with road transport. A single low-value consignment had to be booked in, loaded into a wagon, transported to a transshipment depot, transferred to another wagon possibly as its sole cargo, this would then be taken perhaps to yet another yard before being attached to the local 'pick-up' goods train to be dropped off at the nearest station to its destination. The revenue on the cargo probably only paid for the time of the booking clerk. The 1888 Railway and Canal Traffic Act changed the law so that the Government could place a ceiling on the rates charged by the railways. All companies were required to submit their rates for review at a public hearing and all rates had to be listed in a Rate Book, which was to be made available to the public.

In 1893 the new government ordered railway rates came into force and the railways naturally increased the charges to the maximum allowed in order to recover the losses on enforced discounts. There was a public outcry at the time and another Royal Commission on Railways was hastily set up to review the law. This resulted in more legislation, further restricting the freedom of the railways in altering their charges and leaving them no room for competition other than in the additional services and facilities they could provide.

Constrained by legislation on charging, British main line railway companies began a period of intense competition based on providing better facilities for the customers. Smaller lines could not operate successfully under these conditions and many were purchased or leased by larger concerns. The larger companies were also running into difficulties and dividends were reduced to shareholders. The Midland Railway had improved conditions for third class passengers some years before and this had to be matched by other companies, resulting in an average threefold increase in the weight of train per passenger carried. There was fierce competition for goods traffic, even if it made a loss there would still be some contribution from each consignment toward the fixed costs of operating the railway.

The Government control over railway rates, which were still based upon the value of the cargo carried, coupled with the obligation to carry any goods offered, became a serious problem for the railways with the growth of the petrol driven road lorry traffic in the late 1920s. Road haulage was not governed by the same 'common carriers' legislation, so a road haulier could offer to carry a high value cargo at a lower cost than the railway, leaving the railway to haul back the low value 'empties'. The Government had allowed an increase in railway freight charges in 1920 but then prevented any further rise until 1937. By the mid-1930s competition from road transport was becoming a serious problem for the railway companies, who launched a 'fair deal' publicity campaign. The following selected pages are from the LMS Magazine of March 1932 which illustrate the 'Fair Play for the Railways' campaign and an article titled 'Canals and Transport Costs' which discusses the railway's advantage over canals

It was the Transport Act of 1953 which first substantially altered the legislation relating to 'common carriers' and allowed British Railways to refuse less profitable cargo. Sundries traffic was transferred to National Freight Carriers in 1964 although the railways pulled out of this business completely in 1968. Railway parcels traffic was carried until 1981, after that the Red Star parcels service was operated for some years by British Rail until sold into the private sector in 1999.

Vol. IX   MARCH, 1932   No 3.

EDITOR: O. F. COX, Euston Station

PRINCIPAL CONTENTS

	Page
Tell Your Friends	75
North London Railway. Part II	76
Soap Through the Centuries. Part I	79
Movement of a Circus by Rail, The	82
Nigerian Railway. Part II	84
Canals and Transport Costs	91
Running Tests on Locomotives	92
Departmental, etc., Officers, Directory of	98

Local pages, 87; Ambulance Notes, 97; Children's page, 102; Juniors' page 100; Staff Notes, 95; Temperance Notes, 104; Women's page, 107.

THE question of the unfair competition to which Railways are subject by reason of the unduly favourable circumstances under which road transport operates as compared with rail, is one of great importance not only to ourselves but also to the public and to industry in general. With this in mind, the Railway Companies took the opportunity of a recent meeting with the Minister of Transport in connection with the Final Report of the Royal Commission on Transport, to make certain representations and submit a statement of our case.

Copies of this statement are in the possession of the Company's District Officers, principal Stationmasters, Goods Agents, and others. A pamphlet on the subject entitled "Fair Play for the Railways," has been similarly issued, and for the information of our readers the principal points are clearly and concisely summarised under the title "Tell Your Friends."

INSTANCES of successful efforts on the part of members of the staff to give the public satisfaction and thus attract new business to the Company continue to come to hand. Amongst the latest may be mentioned letters written by influential members of the business community expressing gratitude for the care and consideration shown during the journey of an invalid lady, and appreciation of the courtesy and helpfulness of a ticket collector.

A REAL spirit of co-operation is evinced by an arrangement just completed between L M S executives and the 7,000 men employed at the Carriage and Wagon and Locomotive Works at Derby. In the past, the Derby Works holiday has taken place during August, with a break at Whitsuntide as well; but, as these are peak periods of the year for passenger traffic, the provision within them of the necessary trains for the movement of the Derby railwaymen and their families has become increasingly difficult. The situation was placed frankly before the men's representatives, and as a result of a ballot they have agreed to take their summer vacation from July 16th to 23rd, and for the Whitsuntide break to be transferred to September.

A NEW type of signal which can almost be said to speak is the latest innovation to be introduced by the L M S. It not only tells the driver of a train whether it is safe for him to proceed, but in addition, how fast he may go, and on which route. Officially the new signals are known as "Speed Signals," and are being installed in connection with the widening and development scheme at Mirfield, Yorkshire. The area to be covered extends from Heaton Lodge Junction to Thornhill Junction, a distance of $2\frac{1}{2}$ miles.

"SPEED Signals" are described as being an advance on any other colour light system in use in this country. They consist of coloured electrically-operated lamps whose powerful beam can be seen for 1,000 yards in daylight and for a considerable distance in thick fog. Three light units are fixed one above the other at heights of 8 ft. 6 in., 12 ft. 6 in., and 16 ft. 6 in. from rail level. When a train is signalled on the "high speed" route, the top light is operated, the middle light controls the "medium speed," and the lowest light the "low speed" route. By means of track circuiting, the train automatically protects itself by keeping the signal behind it at danger until it has cleared the next signal ahead. Furthermore, the signalman is always aware of the position of a particular train, as this is indicated by means of lights on a diagram of the track, examples of which are being installed in the new signal-boxes at Mirfield. In addition, the normal aspect of each signal is automatically repeated in the signal cabin.

TELL YOUR FRIENDS

Rail v. Road Considerable interest has recently been focused on the question of rail and road transport. What is it that the railways ask? They urge that legislation should be introduced to give them a fair deal. At present road transport is unduly favoured. The railways had to buy land, build their track, and maintain it; they pay, too, for policing and signalling it. Road users should similarly pay their full share of the cost of new road construction, improvement, maintenance, signalling, and policing.

Railways are bound by Parliamentary regulations; road haulage should also be regulated.

Railways and the Community The railways have been largely responsible for building up the industries of the country. For nearly one hundred years they have been the veritable life-blood of commerce. *They are no less necessary to-day.*

In 1931, for instance, 230 million tons of coal and minerals were conveyed by rail. Over 1,200 million passengers are carried in a year, whilst every weekday at least 200,000 passengers use the London main line termini in a single rush hour.

These are services which road transport could not hope to provide and which show how vital the railways still are to the well-being of the community.

2¼ Million Motors The two and a quarter million motor vehicles which to-day use the road make it necessary to expend the huge sum of £60,000,000 every year for road upkeep, extension, and improvement. Of this sum, the local rates—and the railways themselves are large ratepayers—provide two-thirds, or £40,000,000 a year. The railways claim that all road users should pay their share of expenditure on roads in proportion to the use they make of them. If this were so, the ratepayer would be relieved of the bulk of his £40,000,000 annual contribution to the road budget and the railways would be able to compete with the road on a fair basis.

British Industry first! Every year the railways consume over 13,000,000 tons of British coal as locomotive fuel, the production and handling of which gives employment to 60,000 miners. Motor vehicles use imported oil as fuel, at a cost of £10,000,000 annually. The fortunes of the country depend upon the restoration of a favourable balance of trade. Tell your friends that by using the railways they are supporting British industry and helping to restore the prosperity we all so earnestly desire. Money paid for imported oil goes to the wrong side of the nation's trading account.

What Railways have lost Compared with 1924, it is estimated that no less than £16,000,000 net revenue was lost to the railways by road competition in 1930, even after due allowance has been made for the effect of bad trade. But for this huge loss of traffic, the railways would, in most of the years since amalgamation, have earned more than the profit fixed by Parliament. Your friends would then have benefited, because four-fifths of the surplus would have been used to reduce railway rates and fares; but as a matter of fact, so far from this being possible, an increase of about 7% had to be made in freight traffic rates in 1927, which would have been unnecessary but for road competition. Further increases in rates on coal and other heavy traffic will be inevitable unless the railways are placed in a position to compete with the road industry on equal terms.

Vital to You The whole question is one of vital importance to every railwayman. It means that every railwayman must exert himself to the utmost in the fight against unfair road competition, and spread the facts among his friends so that a widespread interest in the railways' position may be roused all over the country.

Grand National and Easter The two big travel events of the month are the Grand National and Easter. The L M S is proud of the fact that it carries more people to "The National" than any other transport medium. Make this year's traffic a record! Get to know what your local facilities are and suggest to your friends that everyone should see this grand spectacle at least once in a lifetime.

Easter comes early this year, and, coupled with the fact that there has, unfortunately, been a lot of illness this winter, more people than ever should be wanting to get away for a long "pick-me-up" week-end. It is up to us to get busy. Tell Your Friends about L M S "penny-a-mile" Easter travel, about the comfort and the speed of trains as compared with chars-a-bancs.

CANALS AND TRANSPORT COSTS

By E. T. Good

"AS a result of the efforts of British industries to reduce costs to the level of those of competing countries, a great revival is taking place in the use of canals . . . more use being made of England's waterways to meet the demand for cheap and yet speedy transport. . . . The cost of transport by canal is one quarter the amount charged for goods sent by rail."—*A recent Press report.*

* * *

Let us test this theory. It is important, for there is a movement to spend huge amounts of money on canal modernisations and extensions. We are invited to consider what canals do for the trade of foreign countries. But British and foreign conditions differ widely.

Where Canals Succeed

Canals can only succeed in countries, or districts, where the ground is flat, or nearly flat, and where there are navigable natural rivers to link up with the artificial waterways. Nearly all parts of mining and manufacturing Britain are hilly; our levels are bad; our rivers are short, narrow, and shallow, and full of weirs. Foreign canals are nothing like so economical as represented by enthusiastic advocates of inland water transport. Towards the end of February, 1931, it was reported from France that "rail transport was normal, but, owing to floods, canal and river transport was very slow. Some coal-laden barges, despatched about the middle of October, from the Nord and Pas-de-Calais to Paris, for instance, had not reached their destination"—in February! Nearly all through the winter of 1929-30 the canals of Northern Germany, Poland, and parts of Belgium and France, were frozen up. Mines and works dependent on inland water transport were laid idle. Ice in winter, floods in wet weather, and shortage of water in dry weather, frequently interrupt canal transport.

A Costly Service

Regarding cost, it may be mentioned that the canals of France are maintained at the public expense, the Government not only subsidising some of the waterways, but compelling railways to charge extra high rates on coal, etc., in order to enable the canals to compete. Even so, a recent Census shows that the number of canal barges and the number of bargemen have decreased since 1912, notwithstanding the fact that many miles of canals have been added to France by the reversion of Lorraine. Traffic on the railways has greatly increased. In the unlevel parts of Germany it costs more to carry coal by canal than by rail. In the United States a great canal, constructed and maintained at the public cost, has been reported by special Commission as a hopeless failure, but a railway serving the same district has been a success in every respect.

Some Object Lessons

In our own country we have striking object lessons. Not one ton of coal in every twenty received by London from the pits is carried by canal barge, though the canals were constructed long before the railways, though the land is comparatively level between, say, the mines of Nottinghamshire and some other Midland counties and London, and though the rail rates are alleged to be excessive. The railways give better service than the canals give or can give, even in the case of coal, a commodity specially mentioned as suitable for water transport. In the London area there are nearly 200 coal stations or depots, so arranged that only a short cart journey is necessary to take coal from the railway to the householder, besides which the railways maintain countless wait-order sidings in the colliery districts and near the city, for the convenience of the trade. No system of canals can give such facilities or economies.

Some Advantages of the Rail

In Yorkshire we have an extremely fine canal system between the mines and the Humber. Much of the ground is more level than in other British districts; we have the rivers Aire, Calder, and Ouse to assist the transport; and from Goole to Hull there is tidewater—the deep and wide River Humber; nevertheless, for every ton of coal Hull gets by barge it receives between 15 and 20 tons by railway. That fact speaks for itself. Canal transport is not economical, as a general proposition. Only in very special cases, in very selected and limited areas, where the natural conditions afford flat, or nearly flat, ground and good rivers, can inland water transport be successful. Even in such favourable circumstances canals can only serve a minority of traders. We have no pit or furnace or factory or big wholesale warehouse or shipping dock that is not served closely by a railway; but very few industrial or trading establishments are near canals. Not one British mine in six has a canal near-by. Not one factory in ten is near to a canal. But the railways are "on the job," and they can operate practically regardless of floods, shortage of rain, or extreme frosts.

Midland Railway 2000 class 0-6-4T 'Flatiron'

The driver and fireman of 'Flatiron' No. 2004 pose for the camera at Derby while undertaking station pilot duties sometime between 1935 and 1938. Photo: A. W. V. Mace © Transport Treasury

The Midland Railway 2000 Class was a class of 40 0-6-4T steam locomotives designed by Richard Deeley. They were known as FLATIRON or HOLE-IN-THE-WALL tanks because of their distinctive shape; their side tanks extended to the front of the smokebox and they had a cut-out in the side tanks to access the motion.

They were originally developed from 0-4-4T types designed for commuter work, they were numbered 2000–2039. Acceleration and stability were poor, they were rough riders at speed and were liable to oscillate on sub-standard track, this led to a number of derailments after which they were relegated to freight work. All were rebuilt with a Belpaire firebox and superheater between 1920 and 1926. The superheated engines had slightly longer smokeboxes which extended in front of the side tanks. They kept their Midland Railway numbers after the formation of the LMS who gave them the power classification 3P. All were withdrawn between 1935 and 1938. The standard parts would have mostly been used for spares rather than scrap.

Midland Railway 2000 Class 0-6-4T Specifications			
Designer	Richard Deeley	Coal capacity	3½ tons
Built	Derby – 1907	Water capacity	2,240 gallons
Total built	40 locomotives	Boiler	Midland Railway type H1
Driving wheels	5ft. 7in.	Boiler pressure	175 psi
Trailing wheels	3ft. 1in.	Cylinders	Two inside
Length	40ft. 4½in.	Cylinder size	18½in. x 26in.
Weight	72 tons 8 cwt	Tractive effort	19,756 lbs

THE BEYER-LJUNGSTROM TURBINE LOCOMOTIVE

by David Cullen

The 1920s heralded an exciting era for the steam locomotive in Britain. In the pursuit of its ultimate form, engineers began exploring some intriguing technology. This included ultra-high pressure water-tube boilers, advanced valve systems and streamlined locomotive exteriors. Interest was also shown in a potential alternative to cylinder-drive. Operating by converting the 'kinetic' energy of steam in motion into usable 'mechanical' energy, this was the steam turbine. Used for purposes varying from generating electricity in power stations to powering ships, this had existed since the late 19th century in several formats and in many sizes and complexities.

The type relevant here was a 'reaction' turbine of relatively small and simple construction. It comprised a revolving shaft supporting a series of axially-fitted discs termed stages. Taken in the direction of steam-flow, each was of a larger diameter than the previous. These displayed multiple, angled blades similar to a fan. Sealed within a sturdy, gas-tight casing, the arrangement generated rotational drive from high-pressure steam being passed through the blades. This steam expanded considerably throughout the process, producing optimum power as increasing volume during passage acted on the progressively larger stages.

Applied to a railway engine, motion was transmitted via gearing to the driving axle most easily accessible, then to the other driving wheels by coupling rods. The system provided several advantages over a comparable cylinder engine. It obtained a high power output from the considerable kinetic energy possessed by the steam (at 200 lbs./sq. inch, this achieves a velocity equivalent to 4,000 ft./sec. release.) There was simplicity of operation with associated reliability. Power transmission was continuous rather than in pulses, creating smooth running and reducing the likelihood of wheel-slip. With no reciprocating pistons and few whirling masses, in motion the locomotive was less wearing on the track. In addition, a measurable increase in efficiency was normal, with accompanying reductions in fuel and water consumption.

In this country, the London, Midland & Scottish Railway showed keen interest. Completed in 1935 under Chief Mechanical Engineer William Stanier, L.M.S. Pacific No. 6202 *Turbomotive* epitomised the British concept, showing enormous potential in both ability and efficiency *(see MT4, page 10)*. Unfortunately, the turmoil of World War Two resulted in excessive use and inadequate standards of the specialised maintenance required. Deterioration was inevitable, leading to failure of the main turbine in 1947.

Withdrawn by management no longer interested in its potential, the locomotive was placed in store and forgotten. Eventually re-emerging in B.R. days it was rebuilt as a four-cylinder Princess Royal class 4-6-2. Named *Princess Anne* and bearing the number 46202 the engine was pressed into service on the London Euston to Glasgow route in the summer of 1952. Tragically, she was destroyed just months later in the triple-collision at Harrow & Wealdstone station.

While the L.M.S.'s most notable turbine engine, *Turbomotive* was not its first. A unique experimental machine was commissioned in 1926. Constructed by Beyer, Peacock & Co. Ltd. of Manchester and using a turbine supplied by the Swedish company Ljungstrom, while never formally named, logically it became known as the *Beyer-Ljungstrom Turbine Locomotive*. This was of quite extraordinary format.

Virtually every steam locomotive could be defined as a 'tender' engine from carrying its fuel and water supplies in a separate unit, or a 'tank' from storage in containers on its own frames. In a curious hybrid concept, the two-section Beyer-Ljungstrom (B-L) gave the impression of being both.

The leading unit might have been seen as a 'tank', having a recognisable firebox/boiler/smokebox configuration, enclosed cab, rear coal-bunker and side mounted water tanks. This was carried over five sets of wheels. A two-pair bogie-truck supported the front and three individual pairs the section comprising the firebox, cab and coal bunker. At first glance, this gave the impression of a locomotive having the wheel arrangement 4-6-0T. However, of the same modest size, all these wheels were bogies, purely for weight bearing. Creating no motion therefore, this was not a locomotive at all, simply a steam generation plant. This led to perhaps the most unconventional method ever devised for moving a steam locomotive, covered in the following section. Technically it was not a 4-6-0 either. Under the traditional 'Whyte' classification, this referred specifically to a locomotive with four leading bogies and six driving wheels, although benefit of the doubt might be given concerning the '0' reference to trailing wheels.

On initial inspection, the second unit could be taken as a conventional tender. However, this in no way fulfilled a tender's role, carrying neither coal nor water and isolated from the cab by the coal bunker. Noticeably longer than a typical tender, it was mounted over five pairs of wheels. Three large diameter pairs supported the front section and a four-wheeled bogie truck the back. In an arrangement unprecedented and almost beyond comprehension, the three large pairs displayed coupling rods.

The Beyer-Ljungstrom experimental steam turbine locomotive arrives at St. Pancras circa 1926-28.
Photo: Milepost 92½ © Transport Treasury

These were the driving wheels. Following from this, the trailer also accommodated the power turbine. Set in its stout casing, this was mounted laterally across the vehicle's front section along with auxiliary equipment. An absence of shrouding around the front rendered this clearly visible. Further vital equipment was carried aboard; details to follow.

The B-L's overall wheel arrangement has been quoted as 4-6-0+0-6-4 and 4-6-6-4. However, 4-2-2-2-6-4, arguably the most bizarre notation ever put forward, seems the most accurate.

Technical Specifications

Where better to begin than the tender-like unit playing its unique role in the locomotive's operation. Determining the degree of rail-grip, the Weight of Adhesion (WoA) was 54 tons 12 cwt, distributed over the coupled axles as 18 tons 4 cwt, 18 tons 5 cwt and 18 tons 3 cwt from front to rear. Although amassed from bearing large and heavy components, once again, details to follow, this was several tons below the WoA of a typical passenger locomotive. Providing rear-end stability, in particular when running this unit first, the bogie truck carried a further 19 tons 6 cwt. The diameters of the coupled wheels and bogies were 63 and 45 inches respectively. Wheelbase dimensions were: coupled 15ft. and bogie 6ft. 3ins, with a total wheelbase of 27ft. 10⅞ins. Overall truck length was 38ft. 5⅜ins. and height from rail level just over 9ft. Tractive effort was quoted as 38,000 lbs.

The leading unit's wheels were all a modest 39 inches in diameter, giving the boiler and smokebox a massive, powerful appearance. Weight was 69 tons 16 cwt, distributed as 26 tons 2 cwt over the bogie truck and the three individual pairs supporting 14 tons 12 cwt, 14 tons 10 cwt and 14 tons 12 cwt respectively.

Why none was utilised for traction has subsequently raised questions, as not only did this result in lack of hauling ability, it added almost 70 tons to the weight of every train worked. Wheelbase measurements were: bogie truck 6ft. 3ins, individual wheels 14ft. and overall 28ft. 1½ins. Vehicle length was slightly under 38ft. Coal and water stocks were quoted as 6 tons and 600 gallons.

The over-buffers length of both units coupled was close to 80 feet and their total wheelbase just under 63ft. 11ins. Height from rail level to chimney crown was just short of 12ft. 10ins. Chimney diameter was 24 inches. Rail level to the boiler's lateral mid-line was 8ft. 10½ins. Containing 238 tubes, the boiler comprised a barrel of 6ft. 5ins. external diameter with a little under 10 feet between firebox and smokebox tube plates. The top was surmounted by a dome 9¾ inches in height. The firebox was some six feet from grate level to crown and some five feet in length. Grate area was slightly under 30 sq.ft.

Above the firebox were mounted two 2½ inch diameter safety valves set to 300 lbs per sq.in., unusually high for Britain where few locomotives operated above 250. Released by the regulator valve, the steam was raised to 400°F by the superheater, then supplied to the main turbine through a pipe running initially along the right side under the running board. It then traversed beneath the cab to access the rear panel below the coal-bunker. A flexible connector fed the steam to the trailing truck.

The Beyer-Ljungstrom Specialist Components

The main turbine was an 'axial-flow' unit, intended to produce a maximum output of 2,000 h.p. at 10,500 rpm, corresponding to a rail speed of 78 mph. Its hardware comprised a single impulse wheel and eighteen stages, the final stage having double-flow blading. Steam was supplied via an inlet port on the left of the casing as viewed from the unit's left side. Having done its work, it was exhausted via a rectangular duct on the opposite side of the casing. This was larger than the inlet to allow for the high-pressure steam's considerable expansion. The casing comprised two sections bolted together. Using a lifting lug set in the top, the unit could be hoisted in and out and the two sections separated for carrying out inspection, maintenance or repair. Turbine drive was transmitted through three-stage reduction gearing which meshed with a gear wheel on the leading coupled axle. Enclosed within a vaguely coffin-shaped case containing lubricant, the gearing worked at a reduction ratio of 25.2 to 1. (25.2 revolutions of the turbine producing one of the driving wheels) in whole figures 126 producing 5. This equates to one mile of track covered from 8,071/320, taking *pi* as 3.14 in calculating driving wheel circumference. The scheduled route of the B-L was between London and its birthplace of Manchester, approximately 200 miles. This would therefore entail revolutions of 1,614,200/64,000.

A turbine's direction of turn is determined by the slant of its blading. While not impossible, creating contra-rotation is impractical, so if required, reverse drive is usually accomplished alternately. *Turbomotive* was fitted with a separate reversing turbine operating through its own gear train. Disconnected during forward running, this would be manually engaged when required. On the B-L, reverse working was achieved by introducing a component called an 'idler-pinion' into the gear train, creating counter-rotational drive from that stage onward.

There were two specific types of steam-turbine locomotive: 'non-condensing' and 'condensing'. Not surprisingly, 'non-condensing' was the simplest. Locomotives of this type which included *Turbomotive* discharged exhaust steam and created firebox draughting in the conventional manner. The B-L on the other hand, was a 'condensing', this type popular for its major contributions to economy. By maintaining the greatest practical temperature range of the steam, the condensing process created a high degree of thermal efficiency. Instead of exhaust discharging into the atmosphere, in itself wasteful of heat energy, it was fed to a large receptacle in which a vacuum was maintained. The vacuum further advanced efficiency due to the scientific relationships between a gas's pressure, temperature and volume. In an ongoing process, exhaust steam was cooled within this vessel and converted back into water. Creating 'recycling' long before it became a social duty, water would be continuously evaporated, the steam used, exhausted, then condensed and returned to the boiler, each 'unit volume' of water up to eight times. This was intended to cut usage of water and subsequently fuel.

In addition to the driving wheels, main turbine and its gear arrangement, the vehicle accommodated the condenser and auxiliary apparatus; a condensing tank some 28 feet in length, four 6ft. 9in. diameter induction fans for drawing cooling air through it, and a turbine-driven geared pump for returning the water to the boiler. The fans were powered by an auxiliary steam turbine located behind the main turbine, its drive transmitted through two-stage helical reduction gearing and a drive shaft. The condensing hardware was typically bulky and heavy, although as already covered, this weight provided essential traction. The system's main disadvantage was its need for frequent specialist servicing, as indeed did the turbines. Complex and time-consuming, work on these units resulted in 'down time' and loss of revenue being disproportionately high compared with conventional locomotives. This contributed largely to the B-L project's final abandonment.

Condensing locomotives were however common overseas, usually in cylinder format. Their presence was especially notable on the African continent where water was scarce and filling-points few and far between. Here, their low consumption, just one-tenth that of non-condensing machines, was realistic. Accompanied by a typical 10% cut in coal requirement, these factors more than justified heavy servicing costs. In Britain where both resources were plentiful, they were never to gain equal popularity.

A condensing locomotive's *modus operandi* obviously meant there were no exhaust blasts for firebox draughting. The matter was routinely addressed by incorporating a power-driven fan in the smokebox to draw the hot gases through the boiler tubes and maintain steam production. These fans worked well, although constantly exposed to hot cinders, ash and grit, they tended to need frequent replacement. The B-L was fitted with a twelve-blade fan of 28¾ inch diameter driven by an auxiliary turbine located within a somewhat ungainly casing in the centre of the smokebox front. The turbine was supplied with steam through a pipe on the engine's right side. Exhaust was subsequently fed to the condenser via another running along the left. This arrangement led to an archetypal 'condensing' feature. Rather than individual exhaust blasts from the chimney, the engine's motion would produce smoke discharging in a continuous stream accompanied by a gentle whine from the fan.

The smokebox incorporated further intriguing equipment. Set in the front section was apparatus making additional use of its hot gases. Comprising an arrangement of friction-driven rollers, this powered a supply of draughting air to the fire-grate, pre-heated for greater efficiency. Set in a housing on the lower exterior of the smokebox front, the operating mechanism took secondary drive from the fan turbine above.

In Service

Beginning in 1926, the Beyer-Ljungstrom operated the route between London St. Pancras and Manchester. Handling mainly passenger trains, it was on occasion seconded to other duties. In action it must have drawn attention and comments in abundance, and many normally knowledgeable fathers must have been stumped by children's queries about its appearance and workings. Had it proved a success, it might have changed the British steam locomotive forever, or at least the L.M.S. concept. Unfortunately, a number of failings let it down, not uncommon with steam locomotives abandoning established technology. There were, of course, the matters of the leading unit providing no traction and high maintenance costs compounded by excessive revenue loss.

Another major problem never rectified was poor firebox combustion. The mechanical pre-heater was blamed, although in hindsight, replacing the original draughting fan with one larger and more powerful might have remedied this. Designed to produce a 2,000 h.p. output from 10,500 rpm at 78 mph, the main turbine's best recorded figure was 1,650 h.p, corresponding to 8,660 rpm and over 64 mph. This was likely due to inadequate steam production linked to the combustion problem. A further issue was air leakage into the condensing apparatus, weakening the vacuum which reduced thermal efficiency. Subsequently, having only worked until 1928, the B-L was quietly withdrawn and dismantled. Nothing of it remains. Perhaps stemming from the L.M.S.'s overall appraisal, compared with other locomotives little information seems to have been recorded.

In contrast, a Swedish 2-8-0 Ljungstrom turbine locomotive of 1932 proved very successful. Operating on the Grangesberg-Oxelosund Railway on freight duties it showed admirable hauling power and achieved a top speed of 43½ mph. Fuel and water consumptions were reduced by 7¼% and 15% respectively compared with the railway's cylinder locomotives.

So, the B-L experiment was obviously not a resounding success. It is then somewhat ironic that during a visit to Sweden, the L.M.S.'s C.M.E. William Stanier assessed the Ljungstrom turbine 2-8-0 and was greatly impressed. So impressed that he went on to create *Turbomotive* and the potential at least for adopting turbine drive on British locomotives.

The Beyer-Ljungstrom locomotive attracts a large crowd, or maybe they are railway dignitaries, after arrival at St. Pancras station. The locomotive worked on the Midland main line between 1926 and 1928.
PHOTO: MILEPOST 92½ © TRANSPORT TREASURY

LMS MOTIVE POWER DEPOT CODES AFTER 1935
by David Young
(courtesy of Chris Tasker, Manchester Locomotive Society)

As is well known, a new system of engine shed code numbers was introduced by the LMS in 1935. This system was not only continued (with some changes to numbers, as will be seen) by the London Midland Region of BR after nationalisation in 1948, but was also subsequently adopted by other regions.

Prior to 1935, the sheds on the Western A, Western B and Midland Divisions of the LMS had continued to use the former LNWR, LYR and MR numbering systems respectively. The Northern Division, which included the Scottish lines within the LMS network, had no comparable system. For example, Longsight, a former LNWR shed, was coded 16, and had sub-sheds at Stockport, Lees and Altrincham. Patricroft, another former LNWR shed, was coded 34, and had a sub-shed at Farnworth (Plodder Lane). On the ex-LYR lines, Newton Heath was coded 1, Agecroft 13, Bolton 14, Horwich 15 (a sub-shed to Bolton, closing after the Grouping in 1923), Wigan 16 and Bury 20. Finally, on the ex-MR lines, Belle Vue (known as 'Manchester' in 1920) was coded 21, Trafford Park 21A, and Heaton Mersey 21B. Trafford Park and Heaton Mersey were ex-CLC sheds, and their codes related to the ex-MR engines housed there, since these two sheds also housed locos from the former GCR and GNR which came under the jurisdiction of the LNER. Interestingly, the MR had introduced smokebox door shed plates in 1908-09, and this practice was not only continued by the LMS, but also adopted by BR after 1948 and applied to all regions.

LEFT: Stanier Black Five No. 44890 rumbles out of Newton Heath Depot shortly before the end of steam.
PHOTO: MILEPOST 92½ © TRANSPORT TREASURY

In 1935, codes came into use on the LMS which remained unchanged in several instances until the sheds closed. For example, 9A (Longsight) and 9B (Stockport), together with 10D (Plodder Lane), never changed, but others in the area, now a part of Greater Manchester, were changed under BR in later years, mostly by absorption into different districts.

1935 CODES ARE AS FOLLOWS		
Code	Shed	Coding details
10A	Wigan (Springs Branch)	8F from February 1958
10C	Patricroft	26F from February 1958 9H from September 1963
19D	Heaton Mersey	9F from May 1950
19E	Belle Vue	26G from May 1950 26F from October 1954
19G	Trafford Park	9E from May 1950
23D	Wigan (ex-LYR)	Recoded 27D 8P from September 1963
26A	Newton Heath	9D from September 1963
26B	Agecroft	9J from September 1963
26C	Bolton	9K from September 1963
26D	Bury	9M from September 1963
26F	Lees (Oldham)	26E from October 1954 9P from September 1963

Note: 26E had been the code for Bacup prior to the closure of this shed in October 1954.

An undated view at Lees (Oldham) with Stanier 2-cylinder Class 4MT 2-6-4T No. 42551 flanked by a pair of Fowler Class 3 2-6-2Ts, Nos. 40051 (left) and 40056 (right). PHOTO: NEVILLE STEAD COLLECTION © TRANSPORT TREASURY

Reference was made previously to the existence of a small sub-shed at Altrincham, which came under Longsight for administration purposes. The GCR erected a small shed at Altrincham, in which the LNWR had an interest, no doubt by virtue of the joint GCR/LNWR ownership of the MSJ&AR. Five locos and 25 men were out-stationed at Altrincham from Longsight as late as 1925 and it appears that the shed survived until the electrification of the MSJ&AR in 1931.

Under the 1935 scheme, the codes 13A to 13E inclusive had been given to sheds serving the ex-London, Tilbury and Southend lines, plus the ex-North London shed at Devons Road, as all these sheds were part of the LMS network. Shortly after nationalisation, it was decided to incorporate the ex-LTS lines into the Eastern Region of BR and they were given codes 33A to 33C inclusive. The vacant '13' codes were then utilised (briefly) to set up a new District covering the ex-CLC sheds, plus Belle Vue and Lower Ince which had just been taken into the London Midland Region. The short-lived codes were as follows:

13A	Trafford Park
13B	Belle Vue
13C	Heaton Mersey
13D	Northwich
13E	Brunswick
13F	Walton
13G	Wigan (Lower Ince)

By May 1950, the '13' District had been disbanded, and the sheds were given the codes as set out previously, with Northwich becoming 9G, Brunswick 8E, Walton 27E and Wigan (Lower Ince) 10F. In the case of Lower Ince, this code was relatively short-lived, as the shed closed in March 1952.

Another short-lived change took place between December 1956 and February 1958, when Heaton Mersey (9F) became 17E (under Derby – 17A) and Trafford Park (9E) became 17F. This was rather reminiscent of the earlier period when these two sheds came under Sheffield (Grimesthorpe) so far as their LMS locos were concerned. It would have been much more convenient to place them under Longsight, one would have thought.

Reference

For assistance in the preparation of this article, I am indebted to the following publications:

Engines of the LMS – P. Rowledge (OPC 1975).

British Railways Sheds and Codes 1948-71 (Midland Railway Society).

BR Standard Class 5MT 4-6-0 No. 73140 at Patricroft MPD on 22nd August 1966. Photo: Robert Anderson © Transport Treasury

Fowler Class 4MT 2-6-4T No. 42374 at Trafford Park MPD on 15th September 1962. Photo: Robert Anderson © Transport Treasury

Ex-L&YR Class P1 No. 11113 at Agecroft Shed in an undated view. PHOTO: NEVILLE STEAD COLLECTION © TRANSPORT TREASURY

The scene at Bolton shed on 18th May 1968. PHOTO: © TRANSPORT TREASURY

Fowler Class 4F No. 44222 at Wigan Shed on 23rd March 1963. PHOTO: NEVILLE STEAD COLLECTION © TRANSPORT TREASURY

A view of Agecroft MPD (26B) on 24th March 1962. Photo: Robert Anderson © Transport Treasury

Looking like a 'playworn' Wrenn model, Stanier 8F 2-8-0 No. 48*25 is near Cemetery Road bridge, Springs Branch shunting in North Sidings. Although the middle number of the loco isn't discernible the cabside rivets show up well. Photo: Alex Mann © Transport Treasury

Stanier 5MT 4-6-0 No. 44949 at Newton Heath MPD on 14th August 1967.
Photo: Robert Anderson © Transport Treasury

Trafford Park Shed, Manchester after closure with former LMS Stanier 8F 2-8-0s and 2-6-4Ts awaiting removal for scrapping.
PHOTO: COLIN GARRATT © TRANSPORT TREASURY

MIDLAND TIMES • ISSUE 5

THE MIDLAND RAILWAY'S SWANSONG
by Philip Hellawell

26TH APRIL 1962 • Stanier Black 5 No. 45239 at Manchester Central, the Midland Railway's terminus in the city. PHOTO: © TRANSPORT TREASURY

In the Beginning

The birth of the Midland Railway was humble, for years confining its operations to the region around which it was based. By amalgamation it became associated with one of the most notorious figures in the history of railway 'power-broking' and finance, finally emerging into a progressive and prosperous railway company.

The founding father of the Midland Counties Railway was William Stenson, a mining engineer of Coleorton, Leicestershire who, concerned that due to poor transport facilities his county's collieries were losing out to the coalfields of Nottinghamshire and Derbyshire, undertook a preliminary study of the land between Swannington and Leicester. He discussed his proposals with like-minded John Ellis who contacted George Stephenson, to report on the project. His findings were favourable but Stephenson was working on other projects, so he recommended that his son, Robert, be appointed engineer-in-chief.

The opening of this short line solved the coal transportation problem for Leicestershire very effectively, but to the disadvantage of the Nottingham coal interests. However, the foundations of the future prosperity of Derby as a railway centre were laid by linking it with Leicestershire in 1840, as well as opening spurs to Rugby and Nottingham.

Other systems in the locality were the North Midland, which wanted to pioneer a railway route between Derby and Leeds, and the Birmingham & Derby Junction Railway. The completion of these two railway systems led to a rate war, with fares and tariffs cut mercilessly until a permanent peace was secured in an effective manner by the amalgamation of the three companies, the Midland Railway being formed on 10th May 1844, resulting in a company with a combined mileage of 181½.

Railway Mania

The first chairman was George Hudson who acquired the sobriquet "Railway King" in 1844 when railway shares were heading for their peak, his deputy being the afore-mentioned John Ellis. With his astute tactics, many of which were of suspect, if not downright illegal, character Hudson secured control over 1,000 miles of lines, and was a dictator of as many new projects as he could muster. Although it had been he who had effectively established the Midland Railway as such, a series of enquiries, launched by the railways he was chairman of, exposed his underhand methods.

However, by then he had embarked on a wholesale policy of expansion and absorption, with the outcome being a strong system with a firm grip on the Midlands counties. Lines spread in many directions and Derby had become the hub of a railway network with working agreements in far-flung areas. Eventually in 1849, Hudson's power vanished, the railway mania bubble well and truly burst, and his descent into bankruptcy was as complete and dramatic as had been his rise. All too predictably, his fall from grace left thousands of small investors facing ruin.

Hudson had left the railway in a poor way, legal costs of the long battle opposing the Great Northern (GN) bills of 1845 and 1846 to reach York were astronomical, too little had been spent on maintenance, whilst capital for branches to Lincoln and Peterborough was still being paid off and the Midland was hemmed in by rivals GN and London North Western (LNW).

Expansion

John Ellis succeeded George Hudson as Chairman in 1849 and guided the Midland in overcoming the malpractices of Hudson's reign. In this he was aided by the Great Exhibition of 1851 at Crystal Palace which proved to be a turning point for the British economy. Ellis realised that the future lay in becoming a national line and a policy of sustained expansion was born. The MR had designs on London, Manchester, Scotland, and the west of Yorkshire.

Outsmarted by the LNW in acquiring existing lines heading to Manchester, the Midland decided to build its own route, eventually receiving permission to extend its line from Matlock to New Mills. Here it joined the Manchester, Sheffield & Lincolnshire Railway (MS&L) to run into Manchester London Road. In 1865 the Midland, along with the MS&L and the GN, had become a partner of the Cheshire Lines Committee, which opened Manchester Central station in 1880, whereupon, the MR switched most of its trains to Central.

The Midland, however, suffered most severely in regard to its traffic to and from the South where its outpost was at Rugby. The MR could not control its own trade beyond Rugby, thus traffic to the south-east, and above all London, was at the mercy of the LNW, to whom it was effectively handed over.

However, early in 1858 through trains on the Midland began running to Hitchin and over the Great Northern line to King's Cross as a result of which its previous route via Rugby into Euston via the LNW fell into disuse. Not surprisingly though, the Midland found their secondary status at King's Cross irritating as well since they could not always run trains exactly when they wished. Furthermore, Hitchin signal box was instructed to always give precedence to GN trains.

The Midland therefore decided to seek powers to build a new line connecting with Leicester through Bedford. Accordingly, on 22nd June 1863, the Midland Railway (Extension to London) Bill was passed. The route from Bedford took in Luton, St. Albans, and Hendon, reaching London by curving around Hampstead Heath.

25TH MARCH 1975 • Sir George Gilbert Scott's magnificent Grand Hotel building stands proudly on Euston Road adjoining the Midland Railway's St. Pancras station.
PHOTO: G. H. TAYLOR © TRANSPORT TREASURY

Finding the best location for a London terminus was a major challenge but, eventually, St. Pancras between King's Cross and Euston was settled upon. However, bridging the Regent's Canal necessitated the platforms being elevated well above ground level, this problem being turned to advantage by the creation of a huge ground floor cellar area, which fortuitously proved ideal to store hundreds of barrels of beer from the Midlands town of Burton-on-Trent and today houses a thriving retail zone.

The resulting station, arguably London's finest, and one of the world's most attractive, was designed by William Henry Barlow (1812-1902). Main contractors were Waring Brothers, supply and erection of the ironwork being sub-contracted to The Butterley Company of Ripley. Its huge and graceful 690 feet long, 243 feet wide single-arched span roof rising to a height of 100 feet is a railway wonder. After 20 long years the greatest of prizes, a through route to London, had been attained – at a cost estimated to have been of the order of £5 million for the 50-mile extension – roughly double the original budget.

Opening took place on 1st October 1868, the first arrival at 04.15 being the Down mail from Leeds, with the first departure being the 06.15 newspaper train. Up Midland expresses started leaving St. Pancras for the North at 07.45 (in Midland days Up meant in the direction of Derby.) By 1873, the south end of the station was enhanced by the magnificent Midland Grand Hotel, designed by renowned architect Sir George Gilbert Scott in the Gothic Revival style.

1875 saw the Great Western Railway (GW) receive a request from the Somerset and Dorset Railway (S&D) to take over their route, but the GW thought they ought to consult with the London & South Western Railway (LSW) in order to maintain good relations. This gentlemanly conduct brought about its undoing as the LSW then concocted a joint deal with the Midland, behind the back of the GW, to lease the line jointly from the S&D. Unsurprisingly, Sir Daniel Gooch, chairman of the GW, was not impressed by the underhand nature of this move.

The next expansion was seen as northwards to Scotland, which was easier said than done, as the Pennine Chain stood directly in its way. However, the Midland had acquired the (Little) North Western Railway in 1874 which had got as far north as Ingleton, further work northwards having been abandoned on financial grounds. The North Western had decided to concentrate instead on the easier and cheaper route to Lancaster and Morecambe which, with a later extension to Heysham in 1904, gave the MR a rather better route into Belfast than via Barrow-in-Furness.

The problem was that to get to Scotland, Midland traffic needed to be handed over to the LNW which had arrived at Ingleton from the north by taking over the Lancaster & Carlisle Railway. The Midland's station was south of Ingleton, which left the two railways separated by Ingleton viaduct, resulting in

26TH SEPTEMBER 1953 • Fowler ex-LMS Class 4MT 2-6-4T No. 42396 shortly after arrival at Ingleton with a local passenger service.
PHOTO: NEVILLE STEAD COLLECTION © TRANSPORT TREASURY

SEPTEMBER 1987 • The delightful lines of typical Midland Railway architecture at Morecambe Promenade station.
PHOTO: TERRY TRACEY © TRANSPORT TREASURY

through passengers having to walk nearly a mile between the two stations. To counter this, the MR had surveyed a route from Settle to Carlisle, plans for which were approved by an Act of 16th July 1866. In the event, an agreement between the two railways led the Midland to decide to abandon the scheme in 1869. Parliament, though, was not impressed, refused to agree, so build it the Midland had to.

This required the MR to conquer some of the most arduous terrain in British railway engineering. The easiest course available was up the Ribble Valley to Ingleborough, but progress was slow and expensive. Batty Moss necessitated construction of the imposing Ribblehead Viaduct – 24 arches, 105 feet high and 1,328 feet in length. Then came Blea Moor tunnel of 2,640 yards, followed by the summit at Ais Gill, 1,167 feet above sea-level. More tunnels and viaducts were needed to reach Kirkby Stephen and Appleby, the line finally opening for goods traffic throughout on 2nd August 1875 and passengers on 1st May 1876 at a cost of over £3,000,000 (around £290m in today's money) and some 200 lives.

West Riding of Yorkshire

Whilst its extensions to London and Scotland were impressive achievements, the industrial heart of west Yorkshire proved a much more difficult nut to crack. The Midland's main line (inherited from the York and North Midland) by-passed this prosperous part of the West Riding after leaving Royston by going up the Aire Valley to Leeds and then Shipley. The importance of these West Riding towns at that stage is underlined by the fact that the very first regular Pullman passenger service in Britain went into service between Bradford and London on 1st June 1874. The carriages, assembled at Derby from kits made in Detroit, were long by UK standards at 58 feet and also differed by being open saloons with a central gangway.

Not only the MR, but the Great Eastern Railway (GE), with a 1,200-mile network, mainly in East Anglia and once described as "having a great deal of railway, but very little traffic" had its sights set on the West Riding, its problem being that, at no point, was any of its network remotely near a coalfield. Seeing the West Riding, like so many people today, as something of an Eldorado on a distant horizon, the GE spent nearly 20 years trying to get its own line there. Eventually it had to settle for reaching Doncaster by means of a GN joint line and, subsequently, Sheffield.

The Midland, meanwhile, wished to make better use of its Settle-Carlisle line by placing Bradford on a direct, shorter route avoiding reversal in Leeds. At the same time, it was isolated from access to the major manufacturing towns of Dewsbury, Halifax, and Huddersfield as well as a number of smaller towns in an area which it saw as having potentially rich pickings for the company.

11TH JULY 1979 • The 1913-built Midland Railway Type 4c timber signal box at Settle Junction. North of the box, the line for Morecambe diverges to the left, whilst straight on is for Carlisle.
PHOTO: TREVOR DAVIS © TRANSPORT TREASURY

12TH AUGUST 1967 • Stanier Jubilee 6P5F 4-6-0 No. 45593 KOLHAPUR heads north over Ribblehead Viaduct.
PHOTO: NEVILLE STEAD COLLECTION © TRANSPORT TREASURY

It had been a change of mind by George Hudson which put paid to the Leeds & Bradford Railway joining up with the Manchester & Leeds in 1846. A second opportunity for a through route came in 1865 when the Lancashire & Yorkshire (L&Y) sought to link up the railways in Bradford at a cost of £260,000 to be shared with the Midland, GN, and the Leeds, Bradford & Halifax Junction railways. However, the latter three companies declined to co-operate and the L&Y felt the cost too onerous to proceed alone.

1874 had seen an MR proposal for a Huddersfield to Bradford line running through the city to Frizinghall rejected by Parliament, but the seeds of an idea had been sown. So, by 1894, the MR had concluded that it should build a new railway to run from its former North Midland main line at Royston, five miles north of Barnsley, to Bradford to join the former Leeds and Bradford Railway near Forster Square station.

The Midland, guided by its formidable chairman Sir Ernest Paget, thus obtained powers for its West Riding Lines, the route being surveyed in 1896 and authorised by the Midland Railway (West Riding Lines) Act of 25th July 1898. This would cut the distance from St. Pancras to Bradford by 11¼ miles and from London to Scotland by 5¾ miles as well as providing for branches to Huddersfield and Halifax. With an allowed construction time of five years and an estimated costing of £2.1 million, the main line was to be in two sections, the first, just over 8¼ miles long, ending at Thornhill near Dewsbury, and the second, eleven miles long, running from Thornhill to Bradford.

First stop was Dewsbury and, it being axiomatic that the earliest railways got the best routes and the best station locations, the LNW had arrived in Dewsbury just west of the town centre in 1848, the GN had an equally convenient spot on the east side of town from 1874, meanwhile the L&Y had secured the most central location from 1867, although this was not a through station.

By the time the Midland eventually arrived on 3rd March 1906, it had to settle for a location at Savile Town, nearly a mile out of Dewsbury and south of the River Calder, which involved a steep descent through a 188-yard tunnel. The intended passenger station was never built, which would have been on a viaduct with an approach road running down the side of Savile Town Goods Yard at ground level in front of the station. It should be no surprise to learn that it is the LNW station which survives today, on the main Trans-Pennine Express route from Leeds to Manchester, whilst the MR's freight only station closed as long ago as 18th December 1950.

The through line to Bradford would go up the Spen Valley to Oakenshaw then, entering a 2½ mile tunnel, emerge north of Bradford near Manningham. An underground station at Forster Square was part of the plans, along with other stations at Heckmondwike, Liversedge, Cleckheaton and Oakenshaw. With the LNW opening its Heaton Lodge & Wortley New Line up the

Entrance to the Charles Trubshaw designed Bradford Forster Square station – the large building above the station entrance is the Midland Hotel (built 1885). The statue is of factory reformer Richard Oastler (now relocated to Northgate).
Photo: Geoff D. Smith
© Transport Treasury

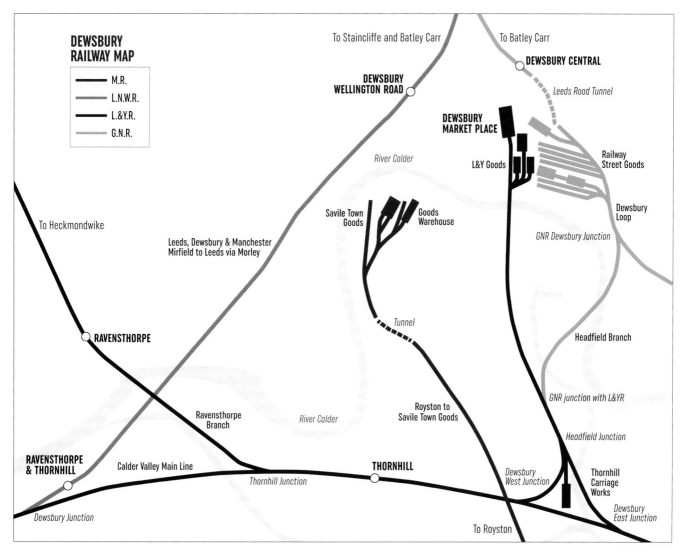

30TH JANUARY 1954 • The driver hangs a wreath on the smoke box door of Tebay's Fowler 2-6-4T No. 42396, about to depart with the last regular passenger train over the Midland Railway's Ingleton branch. PHOTO: © TRANSPORT TREASURY

The Midland Railway's brick-built Sandal & Walton station circa 1912. In 1926 the line was quadrupled, with the new goods lines passing to the east of the two platforms. It closed on 12th June 1961. PHOTO: LENS OF SUTTON © TRANSPORT TREASURY

11TH FEBRUARY 2024 • The imposing 432-yard long 21 arch blue brick Crigglestone viaduct between Royston and Thornhill strides across the Blacker Beck valley. PHOTO: AUTHOR

Spen Valley in 1900, one has to wonder how these moderately sized valley towns could possibly have supported three railway stations.

Nevertheless, striking-off for the West Riding Lines at Royston Junction took place in 1902, with the 8¼ miles to Thornhill being achieved on 10th November 1905. This section contained several noteworthy engineering features, particularly the 432-yard-long blue brick Crigglestone viaduct with 21 arches, and a height of 80 feet crossing the Blacker Beck valley and the similar 307½-yard long Middlestown viaduct of 17 arches, maximum height 50 feet, near Horbury Bridge. There was also one 250-yard tunnel at Crigglestone, lined throughout in blue engineering brick with numerous refuges on both sides.

Opening of the Royston Junction and Thornhill line occurred in 1906 except for the two proposed passenger stations at Middlestown and Crigglestone. Three miles in at Crigglestone two passenger platforms had been constructed in blue brick with appropriate platform edging. They were 450 feet long and 3 feet high above rail level with access ramps down from Haveroid Lane. Identical-sized platforms were provided at Middlestown after 6 miles, but in the event, neither station was ever brought into use except for goods.

At this point, due to the economic climate at the time and the significant engineering challenges which would be needed, the Midland board called a final halt to any work on the route towards Bradford.

October 1967 saw Stanier Black 5 No. 45428 haul the final steam-worked London-bound express service from Bradford to Leeds, before being withdrawn from service a week later after 30 years of service. Following withdrawal this engine was selected for preservation and moved to the North Yorkshire Moors Railway. In 1979 it was named *Eric Treacy* after the renowned railway photographer and former Bishop of Wakefield, since when it has had a couple of major overhauls and has been available for traffic on that line since 2018.

SWANSONG

Unfortunately, this proved to be the Midland's swansong, with its magnificently engineered West Riding Lines enterprise now terminating very modestly at Dewsbury Savile Town goods yard. With the passage of time, the MR's grand plan had lost much of its strategic importance and the unbuilt lines were therefore abandoned in 1911.

The section already built did get some use, though, with goods traffic commencing on 1st March 1906. Passenger traffic began on 1st July 1909 with through trains between Halifax, Bradford, and Sheffield but the Midland had no resource of stock with which to run this service, as its Bradford depot at Manningham was on the wrong side of the city. Consequently, these Midland trains (and, from 1911 some goods trains) were worked throughout in both directions entirely by the L&Y – locos, carriages, and footplate crew.

The line also saw express passenger trains, with a through service from Bradford and Halifax to St. Pancras starting in 1909, using the L&Y's Spen Valley line. Timetables for July 1914 show five Midland trains going to Sheffield with carriages for St. Pancras, variously described as 'Halifax & Bradford' or 'Halifax & Huddersfield', some of which started at Halifax and picked up at Brighouse, Mirfield and Thornhill before heading off onto the Midland system.

Thus, with supreme irony, the only way the Midland could use its high-class, expensively constructed line from Royston to Thornhill was by running trains not from its Bradford Forster Square station but its rival's Exchange station. As can be

5TH AUGUST 1961 • The loco that hauled the final Bradford-St. Pancras express in 1967, Stanier Black 5 4-6-0 No. 45428 at Huddersfield. This engine is now preserved on the North Yorkshire Moors Railway and named ERIC TREACY.
PHOTO: ROBERT ANDERSON © TRANSPORT TREASURY

22ND APRIL 1956 • 1904-built Deeley Class 3F No. 43789 at Royston. PHOTO: NEVILLE STEAD COLLECTION © TRANSPORT TREASURY

22ND APRIL 1956 • 1892-built Johnson Class IP 0-4-4T No. 58066 at Royston shed. PHOTO: J. E. BELL © TRANSPORT TREASURY

28TH FEBRUARY 1959 • 1917-designed Johnson Midland 2F 0-6-0 No. 58260 with 5ft. 3in. driving wheels at Royston MPD. PHOTO: ROBERT ANDERSON © TRANSPORT TREASURY

16TH JULY 1965 • Midland Railway-built Fowler 0-6-0 4F No. 43968 with a short coal train from the Crigglestone direction approaching Royston Junction. PHOTO: MIKE MITCHELL © TRANSPORT TREASURY

FEBRUARY 1965 • A panoramic view of Royston motive power depot, built in 1932 - the last loco shed built by the LMS. PHOTO: WALLY COOPER © TRANSPORT TREASURY

imagined, this left Dewsbury's town leaders very unimpressed at being missed out. One of the oldest fitters at Wakefield shed in the 1960s, Bob Riley, had started in 1920 on the L&Y, and he always said, "You should never trust the Midland."

From March 1925, the LMS introduced the 09.10 Bradford Exchange–St. Pancras and 16.55 return *Yorkshireman* via Low Moor, Cleckheaton Central, and then the Thornhill Junction/Royston line to Sheffield, running time 4¼ hours. For this new working, an all-new set of open carriages (excepting ex-MR clerestory brakes at each end) was provided, with at-seat meals served from a kitchen car. In September 1936, two brand new Stanier Jubilee 4-6-0, Nos. 5724 *Warspite* and 5725 *Repulse*, were allocated to Low Moor from Crewe Works to haul this train. The service continued up to the Second World War, after which it ran from Forster Square via Leeds.

The line was useful for freight, indeed the last engine shed built by the LMS was at Royston, opened in March 1932, which by the 1950s had an allocation of some 50 engines, including a batch of 18 well-maintained Stanier 2-8-0 8Fs. A regular series of these could be seen by-passing Wakefield and hauling coal trains west from the Royston, Cudworth, and Barnsley coalfield areas.

Apart from a brief revival in the early 1960s, all year-round passenger workings on the Royston–Thornhill line had started to fade away in 1946. The eastern end of the line closed in May 1968, but trains continued to use the western section until August, bringing materials to use in construction of the M1 motorway, a final ignominious end to what had been envisaged as a superb arterial main line.

JULY 1962 • No. 45724 WARSPITE, one of the two Stanier 'Jubilee' Class 4-6-0s allocated new to Low Moor in 1936 for the Bradford to St. Pancras YORKSHIREMAN service, see here at Lune Gorge. PHOTO: BARRY RICHARDSON © TRANSPORT TREASURY

Another ignominious end befell the main line from which the Thornhill branch left Royston Junction – surprisingly, an 18½ mile section of the North Midland Railway route from Derby to Leeds (Hunslet Lane) which had been opened on 1st July 1840. To access the Calder valley at Normanton, it had been necessary to excavate the 684-yard Chevet tunnel, the disadvantage of this route being that it missed the major town of Wakefield. However, traffic was so heavy that by 1925 the LMS had to open out this tunnel into a 110ft deep cutting so that a further four miles could be quadrupled to Snydale Junction.

There were 20 daily services each way in the 1930s between Leeds and Sheffield, including the Waverley, Thames-Clyde, and Thames-Forth expresses. However, usage declined over time and, although express trains were still running this way, the Sheffield-Leeds service was down to ten a day in 1960 with Walton station closing on 12th June 1961. Much of the line ran over the Yorkshire coalfield, with attendant subsidence problems, so even though this was shorter, long-distance trains were diverted away to the former Swinton & Knottingley Railway line in 1967. The local Sheffield-Leeds service over this once great trunk route ended on 7th October 1968 and, although they were reinstated in 1973, the through route to Normanton was finally severed in 1986. The single remaining line was used by Bombardier to test and commission the Voyager units for Virgin – 125mph rated tracks (as MML was) of a decent length, and thus why the current depot at Crofton is located where it is. However, the current use of the line is for one train a week to deliver sand to Monk Bretton glass works..

At Royston Junction, it seems surreal that this line (now coded as TJC3 by Network Rail) which passes under the eastern portal of the three arch viaduct is all that is left, when once there were six tracks of main line railway. On a positive note, a 3-mile footpath/cycle way on the Royston-Thornhill trackbed known as the "Chevet Branch Line" was opened in December 2014, surfaced with recycled road materials. The path, through beautiful woodland surroundings, extends from Old Royston to Chapelthorpe and can also be accessed from Newmillerdam country park on the Barnsley Road.

PLANS FOR BRADFORD – DISAPPOINTMENT ALL ROUND

Modifications to the West Riding Lines project, which the Midland had been considering, now began to take shape in 1910 as the Bradford Through Lines Bill. The Spen Valley route of the L&Y was now to be used from Thornhill, where a junction at Oakenshaw would diverge towards Bradford, a distance of five miles, but with a link to connect to Halifax.

Just over a mile from the Oakenshaw Junction would be the southern entrance of a 3,600 yards long tunnel under Bierley Top

31ST MAY 1963 • Clans were occasional visitors to the Bradford area and, here, BR Standard Class 6P5F 4-6-2 No. 72008 CLAN MACLEOD is pictured passing Forster Square box with a stopping service for Carlisle. PHOTO: ROBERT ANDERSON © TRANSPORT TREASURY

and Bowling Park to Ripleyville. From here the alignment was to continue as far as Fairfax Street, where a second short tunnel under Wakefield Road was planned, followed by a cutting on the north side of Diamond Street, leading to a viaduct over Forster Square with high-level platforms 25 feet above the Midland station.

So keen was Bradford Council for this to materialise that they offered the MR rate relief of £8,000 p.a. for 20 years in compensation for the advantages which would accrue to Bradford. As late as 1914 the Midland had purchased Waller's Brewery lying between the passenger and goods lines in Bradford to eliminate a kink where the line was to have passed east of the brewery buildings.

By October 1910, the construction period had been extended on four occasions, although, after being shelved on the outbreak of war the scheme was revived briefly during 1918, but the good years had gone and, on 18th November 1919, the Midland formally abandoned the Bradford Through Lines plan once and for all. Even the Thornhill service was not reinstated until May 1920.

Less than ten miles away, railway development in Leeds had seen that city rise to a position of major regional importance. Leeds had become a major railway crossroads, served by the Midland, GN, NE, L&Y, LNW and Great Central railways providing comprehensive services and connections to all parts of the country.

On 30th September 1920 land in the central area which the MR had acquired for demolition and rail development was sold to Bradford City Council for a big loss-making £295,000. Even so, it seems that BR still had an interest in some property between the two stations in Bradford as late as the mid-1970s, since BR fitter Dave Carter remembers their plant and machinery men being required to service the lifts at properties which stood thereon.

Quite why, despite the austerity of the First World War, the MR board had still felt confident enough to give Bradford City Council a categorical assurance of its intention to complete the Through Lines as late as 1918 is difficult to understand. The writing was on the wall, post-war inflation, declining profits, the growing pre-eminence of Leeds (making any plan to by-pass it irrational), and the mostly quadruple track in place from Sheffield to Skipton were factors which killed off any hope of a railway passing through Bradford.

Abandonment was a staggering blow to the great industrial city of Bradford, as swathes of land in the central area had lain derelict for ten years awaiting development. In fact, it had been prophesied (correctly) by the Yorkshire Observer on 19th November 1906 that "Bradford is today on a siding and there, if this scheme is allowed to lapse, it may remain till the crack of doom." Through passenger traffic would undoubtedly have benefited Bradford, but the majority of express trains would have continued to run through Leeds.

Due to the failure of these proposals, Bradford is left with truncated versions of two dead-end stations facing each other across a gap of only 300 yards (albeit at an altitude difference of some 70 feet) in the perpetual shadow of Leeds. Major redevelopments in the centre of Bradford in the 1950s, 1980s and early 2000s presented further opportunities for linking the lines, not able to be grasped by the authorities of the day.

Tragically, the city lost out again when the 2014 Northern Powerhouse Rail (NPR) proposal (once known as High Speed 3) was abolished, leaving Bradford out of the picture altogether. The scheme envisaged a new high speed rail line from Liverpool to Warrington joining with HS2 in a tunnel which it would share into Manchester. From there, the line would continue to Leeds via Bradford, improving journey times and frequency between the major towns and cities in the north, and allowing them to compete as one large single economy, rather than individually.

However, in 2021, the Government had a change of mind, curtailing the scheme with a new Integrated Rail Plan for the North and Midlands (IRP). Under the IRP the existing lines to Warrington from Liverpool will be upgraded, then, instead of building a dedicated high speed line to Leeds via Bradford, the scaled back scheme provides only a dedicated high speed rail track from Manchester to the Pennine village of Marsden in West Yorkshire. Here the line will join the existing Trans-Pennine route to Leeds via Huddersfield, now being upgraded. Thus, once again the city of Bradford has been denied the great economic development potential of a through inter-city rail line.

Huddersfield branch – a shadow of the grand plan

The Midland also had ambitious and grandiose plans for Huddersfield, including a large passenger terminus and hotel to be called Huddersfield Newtown, some ½ mile north of the joint L&Y/LNW station in the town centre. In 1907, the MR obtained running powers over the L&Y towards Huddersfield but, seen as a threat in part of their heartland, the LNW would not allow the Midland any progress beyond Heaton Lodge.

23rd March 1963 • Mirfield-based Fowler Class 4MT 2-6-4T No. 42406 at Huddersfield Newtown goods yard.
Photo: Robert Anderson © Transport Treasury

However, by an agreement with the L&Y in June 1907, the Midland was able to build a new 4½-mile branch from Lower Hopton, just outside Mirfield, heading south-west through Deighton, Fartown and Birkby to Huddersfield Newtown. The hoped-for grand expansion did not occur, but the land area was big enough for a very sizeable goods station and yard, equipped with a 10 ton crane, and many sidings. A wooden church was built at Fartown for the spiritual care of the navvies.

The branch was engineered for double track, but only a single line was ever laid, opening as late as 1st November 1910. One of the services it provided was the transport of coal taken from Newtown goods yard into the gas works of Huddersfield Corporation (and then North Eastern Gas Board) on Leeds Road, with coke coming in the opposite direction. For this purpose, the council laid a single set of railway tracks down the middle of Beaumont Street in 1922, crossing both the Bradford and Leeds roads and the Huddersfield Broad Canal.

Trains were powered by two small Andrew Barclay 0-4-0 saddle tank engines with guard rails covering the wheels. Locos numbered 1726 & 1783 (both built 1922) ran until March 1963 (and for a few months afterwards if the replacement diesel was out of action). The service was ironically known locally as The Beaumont Street Flyer. Generally preceded by a flagman, the service ceased in 1947, by which time it is estimated that five million tons of coal had been moved by the railway. The remarkable scene of steam trains running down the middle of unguarded residential streets was then no more.

The First World War had finally put paid to any plans for a passenger station at Newtown and, following the 1923 Grouping, the LMS created a junction linking the existing LNW Huddersfield to Leeds route with the MR's Huddersfield line at Red Doles. This gave a direct link from the main line to Newtown goods yard, thus reducing the access branch to 1.2 miles. The longer section back to Mirfield was then hardly used, other than for storing carriages and wagons, being lifted in 1937.

The Newtown goods branch continued in use until final closure in 1968. Today, only the magnificent 15-arch blue brick Bradley viaduct carrying the line over the Huddersfield Broad Canal and River Colne remains as a monument to the Midland Railway's enterprise. More encouragingly most of the route is now public footpath and makes a very agreeable five-mile walk from Mirfield station.

Halifax – crept in by the back door

It would be wrong to call Halifax and Huddersfield twin towns, but they have many similarities and, indeed, today are joint members of the Calderdale & Huddersfield NHS Healthcare Trust. Hugely prosperous in Victorian times, they had a combined

23RD MARCH 1963 • Built 1933 at Derby, Fowler Class 4MT 2-6-4T No. 42406 is pictured near Red Doles junction on the line to Huddersfield Newtown goods.
PHOTO: ROBERT ANDERSON © TRANSPORT TREASURY

population of 135,000 by 1871 and are only 7½ miles apart. Despite this, they have never acquired a direct rail link, the local topography militating against this – the current Northern "service" runs only hourly and takes 23 minutes.

There were, however, more than a dozen attempts to create a direct line, many of which had the MR's sponsorship. In fact, Halifax was the one major west Yorkshire town which the Midland never got to in its own right (although its nearest passenger station to Wakefield was 3 miles away at Sandal, it did have its own goods yard off Westgate station).

That does not mean to say that it didn't expend considerable resources in trying to get to Halifax. The company made its first unsuccessful attempt in 1865, promoting a line from Barnsley to join the LNW's Kirkburton branch, run into Huddersfield and construct with the LNW a joint extension to Halifax. Similar succeeding attempts in 1866 and 1867 also came to nought.

The Huddersfield & Halifax Railway bill was a joint scheme first proposed in 1866 by the LNW and the Midland to reduce the distance by rail between the towns from 11 miles to only six. The scheme was to commence from a junction 1,000 yards north of Huddersfield railway station terminating at a point 100 yards north of Hope Hall, on the east side of Clare Hall Road, Halifax. This line would cross by a 100 feet high viaduct over the existing L&YR line at Elland. With two branches, the overall cost was estimated at £300,000, the lines passing through: Fartown, Birkby, Fixby, Rastrick, Lindley, Elland, Exley, Siddal, Skircoat, Stainland, and Greetland. The scheme was rejected by Parliament in 1866 "on account of the insufficiency of the estimates" of the overall cost.

The West Yorkshire North & South Union Railway was announced in 1871, diverging from the Midland Railway south of Skipton, linking to the Keighley & Worth Valley Railway at a junction near Oxenhope, then on to Halifax, Elland, Stainland, and Huddersfield. From there, a new line would be built along the Holme Valley through Thongsbridge, New Mill, and Hepworth. After crossing the Sheffield, Ashton-under-Lyne, & Manchester Railway four miles west of Penistone, it would run west of Sheffield and reconnect to the Midland near Dore & Totley Station. The line would have been 39 miles in length but, at an estimated cost of £1 million, the promoters were not persuaded to progress.

The Midland was back in 1872 firstly supporting and then withdrawing support for a Huddersfield – Halifax – Keighley scheme, the latter part of which was provided by the GN with its Queensbury lines network. 1874 brought another initiative with a Huddersfield – Halifax – Bradford scheme, a further attempt to open up a connection from Huddersfield to the Midland's new Settle & Carlisle Railway, again the Bill being rejected in Parliament.

However, hope was renewed in 1882 when the Hull & Barnsley Railway was authorised to build an extension to Huddersfield and Halifax from Hemsworth. This grand scheme was part of a planned route for Midland Anglo-Scottish expresses reaching a new station at George Street, the very centre of the town of Halifax, and continuing over the GN to Keighley. Sadly, the parlous financial state of the H&B meant that it failed to reach Barnsley, so the Halifax idea had to be abandoned.

The conclusion of all this was that the Midland had to settle for running rights up the Spen Valley to the Low Moor west curve and approaching Halifax from the east through Lightcliffe. By 1894, the Midland had been given joint use by the L&Y of one of their goods sheds at Shaw Syke to the south of Halifax station. The adjacent shed just north of that was similarly shared by the L&Y with the LNW.

Sadly, this modest gain represented the height of the Midland Railway's ambitious achievements for this prominent industrial West Riding town.

The author would like to thank Dave Carter and Francis Hellawell for their help in compiling this article.

Bibliography

Lost Railways of South & West Yorkshire (2007) - Gordon Suggitt, Countryside Books (2007).

The Midland Railway, A New History – Roy Williams, David & Charles (1988).

Railway Memories No.13, Huddersfield, Dewsbury & Batley – Robert Anderson, Bellcode Books (2000).

A rare photograph dated June 1910, of a Midland Railway train headed by an Aspinall 4-4-0 locomotive of the Lancashire & Yorkshire Railway with Attock carriages, passing through Crigglestone Midland on the West Riding Lines with a Halifax/Huddersfield express to London due to arrive St.Pancras at 19.20. PHOTO: COURTESY OF AUTHOR

21ST FEBRUARY 2022 • Bradley viaduct where it straddles the Huddersfield Broad canal on the Midland Railway freight line from Mirfield to Huddersfield. PHOTO: AUTHOR

28TH JANUARY 2024 • Grade II-listed Notton Bridge – four tracks went through the two right-hand arches on the North Midland Railway from Derby to Leeds, with the West Riding lines to Thornhill passing through the left-hand arch. PHOTO: AUTHOR

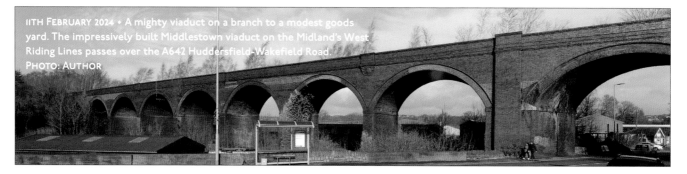

11TH FEBRUARY 2024 • A mighty viaduct on a branch to a modest goods yard. The impressively built Middlestown viaduct on the Midland's West Riding Lines passes over the A642 Huddersfield-Wakefield Road. PHOTO: AUTHOR

COLLISION AT CHAPEL-EN-LE-FRITH
9th February 1957

John Axon G.C. was an engine driver from Stockport (Edgeley) who was killed while trying to stop a runaway freight train on a 1 in 58 gradient at Chapel-en-le-Frith in Derbyshire. This happened after the failure of the steam brake valve led to a complete brake failure of the train engine, ex-LMS Stanier Class 8F 2-8-0 No. 48188.

The collision occurred at around 11.21 a.m. on Saturday 9th February 1957 at Chapel-en-le-Frith (South) on the Buxton to Manchester double line in the Western Division of the London Midland Region. John Axon was in charge of the 11.5 a.m. freight train from Buxton to Arpley (Warrington), comprising 33 loaded wagons and a brake van. The train got out of control as it was descending a steep incline on the Down line. It gathered speed rapidly, and overtook and collided violently with the rear of the 8.45 a.m. freight train from Rowsley to Edgeley (Stockport) comprising 37 wagons and a brake van which was passing through Chapel-en-le-Frith station at about 20 m.p.h.

The initial cause of the accident was the failure of a joint in the pipe leading to the driver's steam brake valve of the Buxton to Arpley train engine. This occurred about one and a half miles from Buxton and not far from the place where the driver would have stopped to pin down wagon brakes before descending the incline to Chapel-en-le-Frith and Whaley Bridge. The fracture not only put the braking system on the engine out of action but also filled the cab with scalding steam. With great bravery and determination the enginemen, after repeated efforts, managed to get the regulator partly closed and to apply the hand brake on the tender, but by this time the front of the train had reached the summit and it could not be stopped. Driver Axon, who was in charge, told his fireman, Ron Scanlon to jump clear and apply the wagon brakes, but in spite of this prompt action the train gathered speed. Driver Axon, however, remained at his post to give warning to the signalman that the train was running away, and in the hope of regaining control on a more favourable gradient. Before this was reached the collision occurred, and I regret to report that Driver Axon and the guard of the Rowsley freight train were killed; the enginemen of this train and the guard of the Buxton train, although badly shaken, were uninjured.

The brake van and the three rear wagons of the Rowsley train were destroyed, and the shock wave through the train derailed four wagons near the front. The engine of the Buxton train fell on its right hand side against the Up platform wall, and its tender struck and demolished the signal box, from which the signalman escaped with a shaking. The leading 30 wagons were piled on top of the rear vehicles of the train in front and formed a mass of debris 25 feet high which blocked both lines at the Buxton end

10TH MARCH 1957 • A month after the fatal accident No. 48188 is pictured at Buxton shed. PHOTO: J. E. BELL © TRANSPORT TREASURY

of the platform. Two of the three remaining vehicles were derailed, leaving only the rear wagon and the brake van undamaged and still on the rails.

The 10.20 a.m. two-coach diesel passenger train from Manchester (London Road) to Buxton, which was standing at the Chapel-en-le-Frith Up platform, was struck by the derailed engine, and the front end of the leading coach was damaged. Fortunately the approach of the runaway train was noticed by the staff in sufficient time for most of the passengers to be taken off the train and none of them were injured. These passengers continued their journey by road about an hour later, and the damaged diesel train was eventually worked back to Longsight Depot under its own power.

The track was badly damaged for more than a quarter of a mile, and the signal box and all its equipment was destroyed. The block and telephone communications from this box were severed, but the telephone from the stationmaster's office was not affected, and he promptly sent calls for ambulances, doctors, fire services and the police, all of whom reached the scene without delay and rendered valuable assistance.

Breakdown trains from Derby and Crewe arrived during the afternoon, but the debris, which included over 150 tons of loose coal was so exceptional that traffic through the station could not be restored until the morning of Tuesday 12th February after an interruption of nearly three days. During this time, passenger trains on the Manchester-Buxton line were terminated at Whaley Bridge, and a bus service was introduced between this station and Buxton. Local freight trains were cancelled and through freight trains were diverted to other routes.

The weather on the day of the accident was reported as fine and clear, and the rails were dry.

In his accident report, Brigadier C. A. Langley wrote: "*This accident was initiated by the very unusual failure of a steam pipe joint which not only put out of action the power brakes on the engine of the Buxton freight train, but also filled the cab with scalding steam at high pressure. This drove the enginemen back from the controls so that they could not shut the steam stop valve nor reach the whistle to give warning of their predicament; they applied the tender hand brake but it was only with great difficulty that they managed to close the regulator partially. Meanwhile the train was still being pushed towards the top of the Buxton incline by the banking engine, the crew of which were unaware of the trouble ahead. By the time this engine had gradually stopped pushing, the front of the train had almost reached the short level section at the summit, and the trains' momentum, together with the power from its engine working with a partially opened regulator, was sufficient to carry it on to the falling gradient where the tender and guard's hand brakes had little effect.*

It is estimated that the train passed through Dove Holes station at about 25 m.p.h. and reached a speed of 55 m.p.h. at the moment of the collision; the speed would have risen to nearly 80 m.p.h. at Whaley Bridge if the train had had a clear run, though it might easily have become derailed on one of the many sharp curves which impose a restriction of 50 m.p.h. between Chapel-en-le-Frith and Disley. If the train had remained on the rails and had not met any other obstruction it would probably have stopped about half a mile short of the second long falling gradient at Disley.

None of the freight train or banking engine crews was in any way to blame for the runaway. Driver Axon and Fireman Scanlon showed great courage and determination in their endeavours to close the regulator after the steam joint had failed, and Scanlon made a plucky effort to carry out his driver's orders to pin down wagon brakes. Driver Axon would have saved his life by abandoning the engine, but he stayed on the footplate to give a warning to the Dove Holes signalman that the train was running away and no doubt in the hope of regaining control on a more favourable gradient. He set an outstanding example of devotion to duty, and I am very pleased to record that he has since been awarded the George Cross posthumously."

John Axon was posthumously awarded the George Cross on 7th May 1957, which was donated by his family to the National Railway Museum, York in 1978. He was also awarded the Order of Industrial Heroism by the Daily Herald newspaper.

Axon was the subject of a 1958 radio ballad (*The Ballad of John Axon*), which was written by Ewan MacColl and Peggy Seeger and produced by Charles Parker.

The full Ministry of Transport report is available to download: https://www.railwaysarchive.co.uk/documents/MoT_Chapel1957.pdf

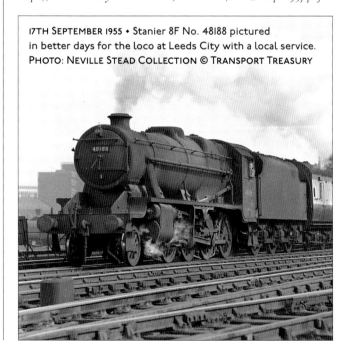

17TH SEPTEMBER 1955 • Stanier 8F No. 48188 pictured in better days for the loco at Leeds City with a local service.
PHOTO: NEVILLE STEAD COLLECTION © TRANSPORT TREASURY

With snow on the distant hills, we see two undated views of Stanier 8F No. 48278 entering Chapel-en-le-Frith (South) from Dove Holes. In the top picture the train has come to a halt while the bottom picture shows the train running wrong-line, presumably with the presence of a flag in the Down loop engineering work is taking place. PHOTO: E. NORMAN KNEALE © TRANSPORT TREASURY

BLACKPOOL'S RAILWAYS

The first railway from Preston to the Fylde Peninsula was opened by the Preston and Wyre Railway in 1840. It went to Fleetwood, via Kirkham and Poulton-le-Fylde. This aided the development of Fleetwood as a resort, as a ferry port and as a fishing port. In 1846 a branch was constructed from Poulton to a station which became known as Blackpool Talbot Road (renamed Blackpool North in 1932).

The Blackpool line joined the Fleetwood line via a severe curve and, following a serious derailment on 1st July 1893, the layout at Poulton was completely redesigned to improve the alignment of the Blackpool route. Another branch opened in 1846, from Kirkham to Lytham. By 1874, the separate railway from Lytham to Blackpool had been joined to the Kirkham to Lytham branch, providing a roundabout route to Blackpool following the coast which ended at a well-situated terminus appropriately called Blackpool Central.

The growth of tourist traffic led to four running lines being provided from Kirkham North Junction to Preston, paired by use, in 1889. Even on the four track sections, congestion was a problem and additional signal boxes were installed with running signals and either no points or just emergency trailing crossovers as a means of 'shortening the block', allowing trains to follow one another more closely. The circuitous route from Kirkham North Junction to Blackpool Central limited capacity, so a 'New Line' was built in 1903, running in a virtually straight line from Kirkham North Junction to a junction at Blackpool South from where four running lines were provided to Blackpool Central. Four running lines were also provided through Poulton-le-Fylde station.

Blackpool to Preston in recent times

As car and coach travel grew, the demand for trains reduced. The impressive railway infrastructure which had allowed large numbers of passengers to be transported was either reduced or abandoned. The direct line from Kirkham North Junction to Blackpool Central was abandoned (apart from around one mile at the Kirkham end which was retained as a Tip Siding). The M55 motorway was built on the rest of the trackbed.

The branch from Kirkham to Lytham and Blackpool was cut back to Blackpool South in 1970 and, from 1982, a single uninviting windswept platform has terminated a singled line from Kirkham which is operated as a 'long siding'.

In 1966, Fleetwood station was closed but a passenger service continued to operate to Wyre Dock until 1970. The branch from Poulton was then used by freight trains as far as the chemical works at Burn Naze. Today, the whole branch is closed but the line is still extant (albeit overgrown and disused). Since June 2006, there have been plans by Poulton & Wyre Railway Society to restore (and preserve) the passenger rail link towards Fleetwood via Thornton & Cleveleys, as a heritage railway.

The Stations
Blackpool Central

Opened in April 1863 as Hounds Hill, it became Blackpool Central in 1878. As excursion traffic grew the station was enlarged in 1901, platforms 7 to 14 being used solely for the heavy summer traffic. The line into Central was controlled by 13 signal boxes as there were extensive carriage yards at Spen Dyke, Bloomfield Road and Waterloo Road – these had siding space for 40 complete trains but were often full, necessitating stabling at nearby locations such as St. Annes, Ansdell and Fairhaven, Lytham and Kirkham amongst others.

In November 1964, after just over 100 years in use, Blackpool Central was closed, this occurred even after the recommendation of British Railways that it should remain open. Blackpool Corporation successfully lobbied British Railways for the closure so that it could realise the potential redevelopment of the site, including – the ultimate irony – road vehicle parking.

When closed, around 18½ miles of track and 39,000 sleepers were removed from the station site and carriage yards.

Blackpool North

Opened as plain Blackpool in 1846 along with the branch from Poulton, it became Blackpool Talbot Road in 1872. The original station was rebuilt in 1898. The rebuilt station consisted of two parallel train sheds and a terminal building, in Dickson Road between Talbot Road and Queen Street. Platforms 1 to 6 were located in the sheds. In addition, there was effectively a separate station at the east end of Queen Street, with open excursion platforms 7 to 16.

After the reprieve of closure in 1964 the station was to be rebuilt in 1974. The train shed was demolished and the excursion platforms then became the new terminus with the 1938-built platform canopy being refurbished to become the new station building.

Blackpool South

Opened in 1903 as Waterloo Road, an additional island platform was added to serve the coast line towards Lytham in 1916. The site was also converted into a junction which connected to the Marton Line, and as with the other Blackpool stations there were extensive sidings to cater for the excursion trains.

The station was renamed Blackpool South on 17th March 1932, the final steam hauled service ran on 3rd August 1968.

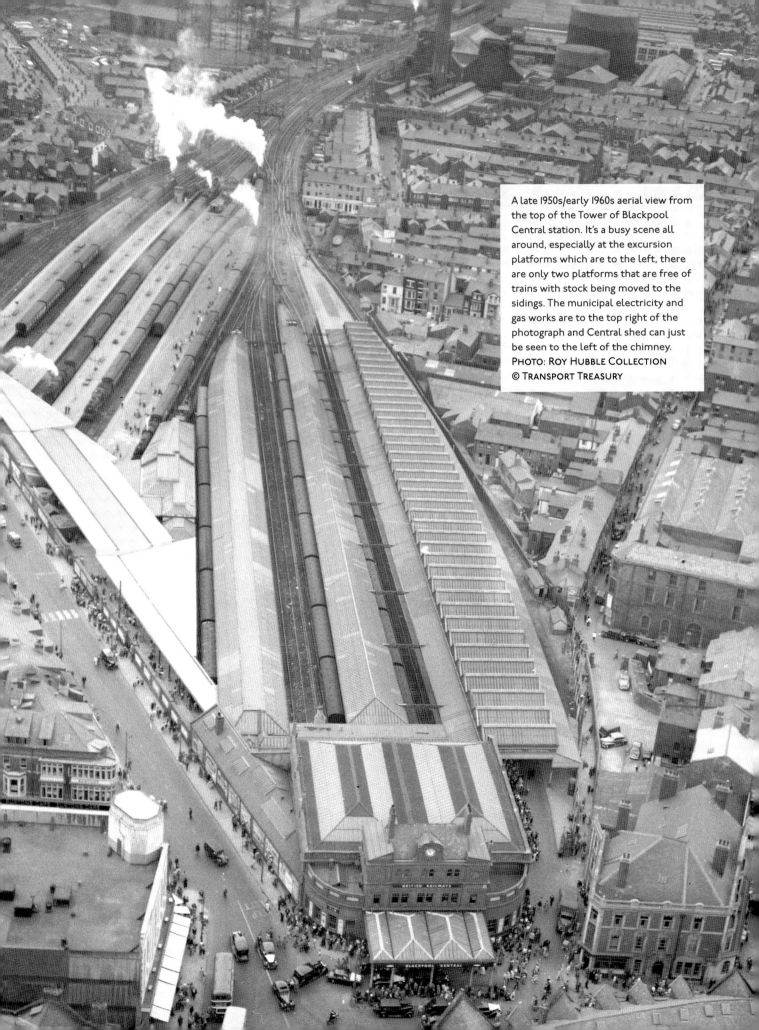

A late 1950s/early 1960s aerial view from the top of the Tower of Blackpool Central station. It's a busy scene all around, especially at the excursion platforms which are to the left, there are only two platforms that are free of trains with stock being moved to the sidings. The municipal electricity and gas works are to the top right of the photograph and Central shed can just be seen to the left of the chimney.
Photo: Roy Hubble Collection
© Transport Treasury

Blackpool Sheds

The two locomotive sheds, Talbot Road (North) and Central (South), were operationally closely linked. Although originally given their own shedcode (Talbot Road 31, Central 32) they were administered as one. The Central shed was the larger and busier of the two, the District Locomotive Inspector was also based there. A new 65 foot turntable was provided at Central in 1932, Talbot Road's was repaired and repositioned.

In the 1935 LMS recoding Central became 24E in the Accrington District with Talbot Road becoming its sub-shed, neither depot were modernised by the LMS. On 1st June 1950 the sheds were recoded to 28A until 1st April 1952 when they received their final change, becoming 10B until closure. This occurred in 1964, North closed on 10th February, Central on 2nd November with the station. The North shed did however see occasional use for servicing locomotives that had been used on excursion trains.

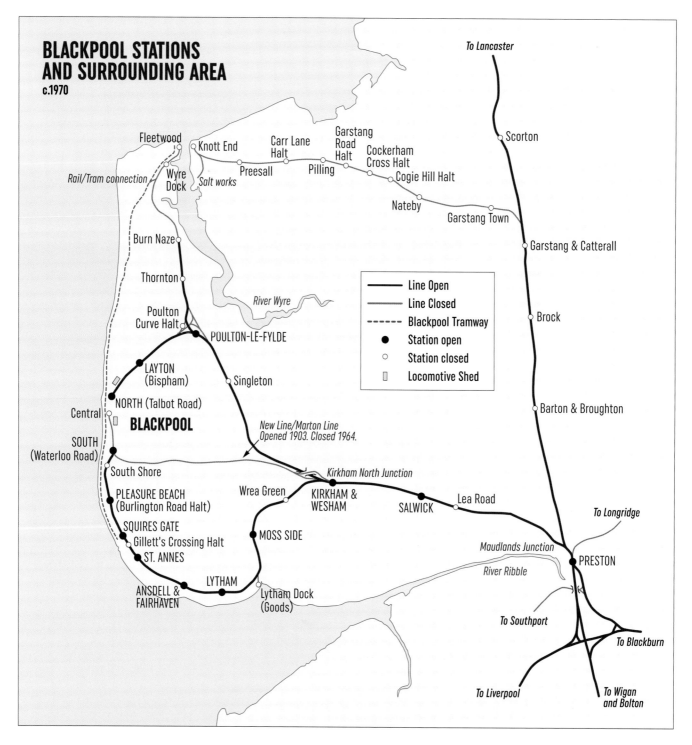

Jubilee 4-6-0 No. 45697 ACHILLES, with Fowler tender, looks immaculate at Blackpool Central, waiting to return holidaymakers to their homes. Date not recorded.
PHOTO: NEVILLE STEAD COLLECTION
© TRANSPORT TREASURY

ABOVE: Derby-built Stanier Class 3MT 2-6-2T No. 40072 pictured at Blackpool Central on 1st January 1960. It was allocated to Central shed from October 1954 until moving to Lower Darwen in December 1961. PHOTO: ALEC SWAIN © TRANSPORT TREASURY

BELOW: The only two identifiable locos on a murky 24th March 1963 at Blackpool Central (South) shed are (left) Stanier Jubilee 4-6-0 No. 45571 SOUTH AFRICA and (right) Black Five No. 44832. PHOTO: ROBERT ANDERSON © TRANSPORT TREASURY

A superb shot at Blackpool Central on 21st August 1964 with Black Five Nos. 45128 and 44798 either side of Stanier 2-cylinder Class 4MT 2-6-4T No. 42625.
PHOTO: LARRY FULLWOOD © TRANSPORT TREASURY

ABOVE: Built by the LMS to a Hughes L&Y design, Class 5P 4-6-0 No. 10455 is seen at Blackpool Central (South) shed on 20th June 1946. Of the seven of the class that went into BR stock it was the only one to be renumbered, 50455. Withdrawal came in October 1951. PHOTO: R. C. RILEY © TRANSPORT TREASURY

BELOW: Aspinall ex-L&Y Class II No. 12430 at Blackpool Central (South) shed on 18th October 1936. Built at Horwich and released to traffic in April 1901 as No. 497, the loco would amass over 50 years in service, being withdrawn by BR on 22nd December 1951. PHOTO: PHILLIP FOX © TRANSPORT TREASURY

A view of Central shed, with the tower prominent in the background, taken on 10th June 1956.
PHOTO: NEVILLE STEAD COLLECTION © TRANSPORT TREASURY

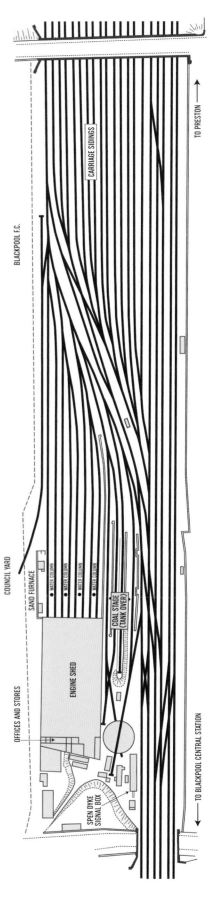

BLACKPOOL CENTRAL (SOUTH) SHED c.1950

ABOVE: Ivatt 2MT 2-6-2T No. 41262 pictured on the approach to Blackpool North station in August 1951. Note the extensive carriage sidings to the left. PHOTO: JIM FLINT & JIM HARBART COLLECTION © TRANSPORT TREASURY

BELOW: Another image taken in August 1951 of a train on the approach to Blackpool North station, this time with Stanier Black Five 4-6-0 No. 44734 of Newton Heath shed in charge. PHOTO: JIM FLINT & JIM HARBART COLLECTION © TRANSPORT TREASURY

BLACKPOOL TALBOT ROAD (NORTH) SHED c.1950

A view of the three-road shed at Blackpool North on 10th June 1956 with Stanier 2-cylinder Class 4MT 2-6-4T No. 42636 in steam. PHOTO: NEVILLE STEAD COLLECTION © TRANSPORT TREASURY

ABOVE: Another view of Aspinall ex-L&Y Class II No. 12430, this time the loco is pictured in the North yard on 20th June 1946. Note that the smokebox number plate has been removed.
PHOTO: R. C. RILEY © TRANSPORT TREASURY

BELOW: Aspinall ex-L&Y Class 1008 2-4-2T No. 50757 at Blackpool North in May 1950 with its former owner's insignia still prominent on the tank. Coupled behind is Class II 0-6-0 No. 52415. PHOTO: JIM FLINT & JIM HARBART COLLECTION © TRANSPORT TREASURY

ABOVE: Passing the watching gallery, ex-L&Y Class 1008 2-4-2T No. 10878 at Blackpool South in August 1937. To service in June 1905, the loco was withdrawn by the LMS in May 1946. PHOTO: HARRY LUFF, ONLINE TRANSPORT ARCHIVE © TRANSPORT TREASURY

BELOW: Black Five No. 45388 at Blackpool South on 20th July 1968 with IP58, the 20.48 Preston portion of the 17.05 Euston to Carlisle service. The loco crew and station staff seem to be having a good chat before departure. PHOTO: © TRANSPORT TREASURY

49509 AT HUDDERSFIELD

Fowler Class 7F 0-8-0 No. 49509 is about to rattle through Huddersfield station with a lengthy freight on 12th March 1952. Built for the LMS at Crewe Works and put into service on 2nd May 1929. It was allocated to Lees (Oldham) shed (26F) until moving to Agecroft (26B) in December 1957, from where it was withdrawn on 16th May 1959.

Photo: © Transport Treasury

FAR REACHES OF THE LMS
Words and Photos by Alan Postlethwaite

This article completes the presentation of my BR(M) photos from 1959-67. They were taken on trips to other places rather than as part of a dedicated plan to capture the LMS. The Highland set was taken during my honeymoon, staying at Inverness with trips to Aviemore (to go skating) and to Kyle of Lochalsh in a blizzard, enjoying toasted tea cakes in the Griddle Car. My previous articles covered the Stafford area (MT2), the Knotty (MT3) and the Midland Railway (MT4).

On a murky winter's morning in 1962 on the LNWR a few miles west of Coventry, Stanier Black Five No. 45276 heads a short parcels train. The junction to Nuneaton is guarded by a fine LMS right-hand signal bracket, so tall that it needs three support guys. The cutting is exceptionally well manicured.

LEFT: Nuneaton Up Signals is a contender for Britain's most attractive signal box. This LNWR Type 4 is (variegated) brick-built to the operating floor and has modest oversailing. Note the loudhailer, external lamp and sun-shade.

BELOW: At Coventry in 1962, work is in progress to rebuild the station prior to the installation of overhead electrification and colour-light signalling. Meanwhile, during the transient diesel phase, a service to Euston is headed by English Electric Type 4 No. D269.

On the joint GWR-LNWR line south of Shrewsbury, Dorrington closed to passengers in 1958. Four years later, the platforms are chamfered off but the yard is still active with general goods and what look like milk wagons. The building is in a restful Italianate style with unusual windows.

Dorrington signalman, Mr E. H. Marsh, is immaculately turned out and all levers and instruments are gleaming. The box is of GWR design with rear catch handles and lever leads which are of engraved ivorine. The block shelf includes various brass release plungers, repeaters, a tapper bell and a time release instrument; this is the item in the large circular glass case. Behind on the window sill are further repeaters this time in wooden cases.

A cast iron warning sign near Shepton Mallet High Street where the GWR crossed the S&D.

Standard class 3 tank No. 82005 heads a stopping train on the joint GWR-LNWR line south of Shrewsbury. The shallow cutting and hedgerow-fences are beautifully manicured, a credit to the gangers.

Condover was a closed station on the joint GWR-LNWR line south of Shrewsbury. The tubular steel signal post is GWR. Wind-bent trees complete this desolate railway landscape.

On the North London line, Canonbury's building could be mistaken for some on the South London line.

In 1959, enthusiasts are taking a break at Canonbury from the RCTS 'London and North Kent' rail tour.

Camden Road Junction was guarded by a North London Railway Type 3a signal box with an extension for power signalling equipment and a generous coal bunker. The quadruple line becomes double here for a short section. The routes beyond both lead to Willesden Junction. The left-hand route leads directly to the LNWR main line via Primrose Hill while the right-hand route takes a more circuitous route through north suburbia.

At Inverness in 1967, BRC&W/Sulzer Bo-Bo Type 2 diesel-electric No. D5346 is fitted with snow ploughs at the wheels. The signal posts are LMS tubular (foreground) and Highland Railway lattice (beyond). The two auxiliary signals are 'Warning' (left) for something unsighted and 'Calling-on' for light engines.

Opening in 1855, the Inverness and Nairn Railway became part of the Inverness and Aberdeen Junction Railway and then part of the Highland Railway (as far as Keith), later absorbed into the LMS. Nairn station was spaciously laid out and was in immaculate condition in 1967. The station building has an integral canopy and a bay window for the waiting room. The stationmaster's house stands beyond in its own garden and there are circular flower beds all along the Down platform. A diesel engine shunts in the goods yard.

A Swindon-built DMU arrives at Nairn where parcels and milk churns are waiting to be loaded. The far platform has two types of barrow. The open lattice footbridge is a common type found all over Britain. Near-identical examples could be found at Cowes, Isle of Wight, and at Gorgie East, Edinburgh.

The LNWR line from Northampton was the first to reach Peterborough, preceding the Great Northern. Oundle station served the nearby public school. Its stone building was in Tudor style with elaborate tiling and clusters of tall chimneys. Note the stone wall, oil lamps and wooden Way Out sign, all probably originals. The building is Grade II listed. The goods yard was still active in 1963.

Other stations on the Nene Valley line were more plain and simple. Barnwell had a rendered brick booking office and a vernacular wooden waiting room with salient canopy. Arriving with a westbound train is ex-LNER class B1 No. 61059. This line made an end-on junction with the Great Eastern at Peterborough East. The line closed under Beeching but the section from Yarwell Junction to Wansford is preserved as the Nene Valley heritage railway.

46246 CITY OF MANCHESTER AT CHEDDINGTON c.1961
Photo: Doug Nicholls

THE PLATFORM END

Picture: The RCTS Hertfordshire Rail Tour No. 2 has reached Stanmore Village on 27th April 1958. The tour commenced at Fenchurch Street at 12.00 noon, returning to Broad Street at 18.55. Photo: Gerald Daniels.

In future issues our aim is to bring you many differing articles about the LMS, its constituent companies and the London Midland Region of British Railways. We hope to have gone some way to achieving that in this issue.

Midland Times welcomes constructive comment from readers either by way of additional information on subjects already published or suggestions for new topics that you would like to see addressed. The size and diversity of the LMS, due to it being comprised of many different companies, each with their differing ways of operating, shows the complexity of the subject and we will endeavour to be as accurate as possible but would appreciate any comments to the contrary.

We want to use these final pages as your platform for comment and discussion, so please feel free to send your comments to: midlandtimes1884@gmail.com or write to:
Midland Times, Transport Treasury Publishing Ltd., 16 Highworth Close, High Wycombe HP13 7PJ.